Small Electric Vehicles

Amelie Ewert · Stephan Schmid · Mascha Brost · Huw Davies · Luc Vinckx
Editors

Small Electric Vehicles

An International View on Light Three- and Four-Wheelers

 Springer

Editors
Amelie Ewert
German Aerospace Center e.V. (DLR)
Stuttgart, Baden-Württemberg, Germany

Stephan Schmid
German Aerospace Center e.V. (DLR)
Stuttgart, Baden-Württemberg, Germany

Mascha Brost
German Aerospace Center e.V. (DLR)
Stuttgart, Baden-Württemberg, Germany

Huw Davies
Institute for Future Transport and Cities
Coventry University
Coventry, UK

Luc Vinckx
Elephant Consult B.V.B.A.
Tervuren, Belgium

ISBN 978-3-030-65845-8   ISBN 978-3-030-65843-4   (eBook)
https://doi.org/10.1007/978-3-030-65843-4

Graphic: DLR, CC-BY 3.0

This Springer imprint is published by the registered company Springer Nature Switzerland AG
The registered company address is: Gewerbestrasse 11, 6330 Cham, Switzerland

Preface

With a growing number of electric vehicles (EVs) worldwide, the EV stock of passenger cars reached 5.1 million in 2018 with battery electric vehicles (BEVs) holding 64 % (IEA 2018). Especially, the sales numbers and models available on the market of larger vehicles, i.e. large cars, SUVs or pick-ups, grew significantly in the past years. This is problematic from an ecological point of view, as they require more energy for operation than small and lightweight electric vehicles (SEVs) and are in most cases a less efficient transport option. SEVs offer the same benefits that come with the deployment of BEVs, such as no emission of exhaust pollutants, but they beyond that require less critical raw materials for the production of batteries and overall emit less greenhouse gases (GHGs) than large vehicles or vehicles with internal combustion engine (ICEs). SEVs are an alternative especially in urban areas. Due to their small size, they occupy less space and could therefore help in the development of attractive city centres. In order to holistically evaluate the potential of SEVs for optimized land use, it is important to consider which transportation modes might be substituted.

While figures of SEVs in China are growing rapidly with 50 million electric three-wheelers and an estimated 5 million low-speed electric vehicles (LSEVs) in 2019, other countries show by far smaller numbers. Especially, considering world markets such as the USA and Europe, SEVs have only limited success. Different rules for homologation complicate a comparison of world markets and the introduction of vehicle models into new markets.

This book should give a first overview of different SEV types of vehicles and their possible applications. It aims for an international view on chances and obstacles for SEVs as well as new research, pilot projects and developments in the area. The present status of SEV technologies, the market situation and main hindering factors for market success as well as options to attain a higher market share including new mobility concepts will be highlighted.

The book is realized against the background of the International Hybrid and Electric Vehicles Technology Collaboration Programme (IEA-HEV) in the framework of the International Energy Agency within the Task 32 on small electric vehicles.

Definition of Small Electric Vehicles The term small electric vehicles comprises in this book three- and four-wheel vehicles which are powered by a locally emission-free drive. Depending on the regional background, they are classified differently, e.g. as low- and medium-speed vehicles, low-speed electric vehicle or kei cars. According to EU regulation (No. 168/2013), they belong to one of the L-categories L2e, L5e-L7e. Additionally, electric cargo bikes and vehicles of categories M1 or N1 which do not exceed 3.5 m, a maximum drive power of 55 kW and an unladen weight of up to 1200 kg, are in the scope.

Review Process The contributions in this book were reviewed by experts in the field in a double-blind two-step process. The editorial team greatly appreciates the reviewers who contributed their knowledge and expertise to the book's editorial process over the past six months. The editors would like to express their sincere gratitude to all reviewers for their cooperation and dedication in 2020 including:

Adrian Braumandl, Karlsruher Institut für Technologie (KIT)
Annick Roetynck, LEVA-EU
Prof. Dr.-Ing. Alexander Müller, Hochschule Esslingen
Christian Ulrich, Deutsches Zentrum für Luft- und Raumfahrt e.V.
Christian Wachter, Deutsches Zentrum für Luft- und Raumfahrt e.V.
Christina Wolking, Technische Universität Berlin
Değer Saygın, SHURA Energy Transition Center
Florian Kleiner
Friederike Pfeifer, IKEM—Institut für Klimaschutz, Energie und Mobilität
Gabriele Grea, Università Bocconi
J. R. Reyes Garcia, University of Twente
Jörg Sonnleitner, Universität Stuttgart
Marc Figuls, FACTUAL
Paul Nieuwenhuis
Prof. Peter Wells, Cardiff University
Dr. Richard Barrett, University of Liverpool in London
Sebastian Sigle, Deutsches Zentrum für Luft- und Raumfahrt e.V.
Susanne Balm, Amsterdam University of Applied Sciences
Sylvia Stieler, IMU Institut
Prof. Dr.-Ing. Volker Blees, Hochschule RheinMain
Werner Kraft, Deutsches Zentrum für Luft- und Raumfahrt e.V.

Stuttgart, Germany	Amelie Ewert
Stuttgart, Germany	Stephan Schmid
Stuttgart, Germany	Mascha Brost
Coventry, UK	Huw Davies
Tervuren, Belgium	Luc Vinckx

Contents

List of Figures

The ELVITEN Project as Promoter of LEVs in Urban Mobility: Focus on the Italian Case of Genoa

Small Electric Vehicles in Commercial Transportation: Empirical Study on Acceptance, Adoption Criteria and Economic and Ecological Impact on a Company Level

**An Energy Efficiency Comparison of Electric Vehicles
for Rural–Urban Logistics**

**Electrification of Urban Three-Wheeler Taxis in Tanzania:
Combining the User's Perspective and Technical Feasibility
Challenges**

Small Electric Vehicles (SEV)—Impacts of an Increasing SEV Fleet on the Electric Load and Grid

Fields of Applications and Transport-Related Potentials of Small Electric Vehicles in Germany

KYBURZ Small Electric Vehicles: A Case Study in Successful Deployment

BICAR—Urban Light Electric Vehicle

Conception and Development of a Last Mile Vehicle for Urban Areas

Development of the Safe Light Regional Vehicle (SLRV):
A Lightweight Vehicle Concept with a Fuel Cell Drivetrain

List of Tables

KYBURZ Small Electric Vehicles: A Case Study in Successful Deployment

BICAR—Urban Light Electric Vehicle

Conception and Development of a Last Mile Vehicle for Urban Areas

Introducing SEVs

Small Electric Vehicles—Benefits and Drawbacks for Sustainable Urban Development

Amelie Ewert, Mascha Brost, and Stephan Schmid

Abstract Small electric vehicles (SEVs) have the potential to contribute to climate protection, efficient land use, and mitigation of air pollution in cities. Even though, they show many benefits that could enhance urban quality of life, they are not yet widely used. In this paper, benefits as well as drawbacks for these vehicles are discussed by combining literature research and outcomes of a mixed-method approach with expert interviews and an online survey. Resulting from these arguments, a vision for SEVs in urban areas is drawn showing them integrated in a mix of various transport modes. Environmental benefits are derived, for example, from their lower weight and low maximum speed making them a more energy-efficient transport option than heavier cars. Additionally, the small vehicle size lowers land use for SEVs and, e.g., allows for less parking areas needed. However, they also hold constraints that need to be dealt with in different ways. On the one hand, the lower safety compared to passenger cars is an issue that is further worsened by current traffic regulations. On the other hand, costs in terms of purchase prices seem to be an issue for SEVs.

Keywords Small electric vehicles · Vehicle concepts · Sustainable transport

1 Introduction

Cities are growing worldwide due to an increasing population, and simultaneously, motorization intensifies. Challenges such as local environmental pollution, a lack of space, and saturation of existing infrastructure are thereby becoming more pressing. The urgency to act and the need for new forms of mobility sets the tone not only for politics, urban, and transport planning but also leads some companies to offer new solutions. One contribution to climate protection and to cope with local challenges

A. Ewert (✉) · M. Brost · S. Schmid
German Aerospace Center (DLR), Institute of Vehicle Concepts,
Pfaffenwaldring 38-40, 70569 Stuttgart, Germany
e-mail: Amelie.Ewert@dlr.de

© The Author(s) 2021   3
A. Ewert et al. (eds.), *Small Electric Vehicles*,
https://doi.org/10.1007/978-3-030-65843-4_1

is the deployment of small and lightweight electric vehicles when replacing heavier cars and being applied together with other ecological transport modes like public transport.

This paper discusses benefits and drawbacks that could result through a more widespread usage of SEVs in cities. Adding to a literature-based research is results from qualitative and quantitative methods in a mixed-method approach. Analyses include expert interviews and an online survey. The results show that these vehicles bring many advantages within urban areas. In addition to other aspects, especially lower land use due to the small vehicle size offers potential by conversion of traffic areas and increased air quality. If SEVs would replace vehicles with internal combustion engines (ICE), significantly fewer air pollutants could be emitted. Due to their lower weight and maximum speed, they are even more energy-efficient than most normal battery electric vehicles (BEV).

Nevertheless, the survey showed that there are many hurdles to be overcome. These drawbacks affect the development of the vehicle technology and transport planning within the cities. These include, for example, safety aspects, e.g., as the vehicles are very light which is often connected to lower passenger safety and crash tests are not required by EU law for type approval of this vehicle category. An example for drawbacks regarding city planning is that most cities are not designed for these vehicles and therefore do not offer advantages in use, such as privileged use of lanes or parking spaces. In a global comparison in some world regions, there is a large market for SEVs such as the Asian countries China, Japan, or India [1]. Europe and the United States, however, only show small sales numbers [2, 3].

In the following, the term SEV will first be narrowed down and explained. Then, advantages and limitations of the vehicles are presented using literature research supported by the results of a qualitative and quantitative survey. For the last section, a future vision of how urban mobility could look like is drawn including all types of mobility.

SEV definition. SEVs in this chapter are referred to three- and four-wheeled L class vehicles according to EU Regulation No. 168/2013. They also include electric vehicles of categories M1 or N1 which do not exceed 3.5 m, a maximum drive power of 55 kW, and an unladen weight of up to 1200 kg.

2 Mixed-Method-Approach

Adding to desk-based research, collecting data and existing literature on benefits and drawbacks of SEVs a mixed-method-approach consisting of quantitative and qualitative empirical social research was carried out. For the qualitative approach, semi-structured expert interviews were conducted. This way, it was possible to derive exclusive knowledge from professionals with different backgrounds by giving insights to practical application, experiences, and research. The evaluation approach is based on a concept of Meuser and Nagel [4] and follows the approach of qualitative content analysis. In a repetitive process, successive categories are formed. The content of the interviews is encoded by paraphrasing individual text passages with the same content. They are classified thematically with categories which are congruent with the key questions. Further, sub-codes comprise partial aspects. Then, statements can be compared and conceptualized. Ultimately, a theory is created by inductively generalizing statements on the basis of individual findings [5]. While the interviews were being conducted, at the same time, the target groups filled out a standardized online survey. Combining these methods in a concurrent triangulation similarities, divergences and additional information could be derived and thus ensure higher validity of information. For both methods the same research questions were applied following a parallel design QUAL + QUAN [6]. The survey took place from March to October 2018 on three main topics:

- Knowledge about SEVs within municipalities and the urban population
- Target groups and usage concepts
- Obstacles and chances for SEVs.

The survey collects assessments of international experts from municipalities, research institutes, consultants, associations, and manufacturers. In total, 32 telephone interviews were held, and the online questionnaire had a sample of 90 with respondents from Asia, USA, and Europe. For both methods, results are not representative due to the limited number of experts.

3 Definition of Small Electric Vehicles

Literature contains a vast array of descriptions, definitions, and categorizations of SEVs and differs regionally. Some of the designations are listed in Table 1 with respective regulations. The categories often include not only type approval for vehicles with electric drive but also vehicles with ICE. In Japan, for example, the term new Mobility vehicles is used which includes Kei cars that have been in use since the 1950s. In this publication, the term SEV is applied as a superordinate term.

European regulation. There are various country-specific categories which in turn have different designations and regulations for type approval. For the vehicles in the scope of this paper, the European regulatory framework for L-category vehicles (L2e, L5e, L6e, and L7e) is defined in Regulation No. 168/2013. Micro- and subcompact electric vehicles with four wheels could in certain cases also be part of category M1, passenger cars, that are laid out in 2007/46/EC. M1 is defined as a light category of motorized vehicles. Some technical parameters of cars in this segment are limited, e.g., number of passenger seats must not exceed 8, but have no further requirements regarding, e.g., mass, maximum speed (minimum speed is 25 km/h), or measurements. Even though there are differences between the M1 category and L7e such as the maximum mass, required crash tests, and width, some vehicles that fulfill requirements of both categories can be registered as either M1 or L7e [8, 11, 12].

Table 1 Alternative terms for SEVs in different countries and limitation of parameters [1, 7–10]

Country	Term	Max. speed (km/h)	Max. mass (kg)	Max. power (kW)	Max. dimensions L × W × H (m)
USA	Low-speed-vehicle (LSV)	40	1360	–	–
	Medium-speed-vehicle (MSV)	56	2268	–	–
	Also: Neighborhood electric vehicle (NEV)				
China	Low-speed electric vehicle (LSEV)	Legislative proposal for vehicles with speed limits <100 km/h in discussion			
Japan	New mobility vehicle	–	–	8	3.4 × 1.48 × 2
	Mini vehicle (Kei car)	–	–	47	3.4 × 1.48 × 2
South Korea	Low-speed electric vehicle (LSEV)	60	1361	–	–
Europe*	L-category vehicle	Various**	Various**	Various**	Various**

* small M1 vehicles are not further specified in this table
** limitation of parameters depending on subcategory

Most of the L-subcategories have the same permitted maximum dimensions: width $\leq$ 2 m, height $\leq$ 2.5 m, and a length that varies between 3.7 and 4 m. The maximum speed varies between 45 km/h, 90 km/h, or no maximum speed. The L2e vehicles are defined by three-wheeled mopeds and represent the lightest class in terms of weight limit ($\leq$ 270 kg). Category L5e includes three-wheeled tricycles, which can weigh up to 1000 kg without batteries. L6e describes light on-road quadricycles with four wheels and a maximum unladen weight of 425 kg. The maximum power permitted varies between 4 and 15 kW, although 15 kW only relates to class L7e. An exception is L5e with no power or speed limitations.

4 Benefits and Obstacles Derived from SEVs

The survey included perception on advantages and disadvantages of the application of more SEVs within urban traffic. The most important aspect mentioned in the interviews is the reduction of land usage especially in regards to stationary traffic. Another major advantage mentioned in the interviews and in the online survey is seen in the vehicles' light weight and the corresponding low energy consumption. An important prospect in particular for municipalities is the improvement of air quality. Air pollution in cities is a concern worsened by population growth and motorization rate. Figure 1 shows the quantitative online evaluation and draws decisive advantages to the reduction of land use and air quality. One benefit that stands out while occurring sporadically in the interviews is noise reduction. An

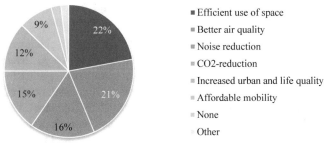

Number of selections 248 (n=84, multiple choice max. 3 ticks)

Fig. 1 Prospects for more SEVs in cities

important component in the decision of transport options for consumers is costs. Compared to EVs these vehicles are less expensive and have lower operation costs. The costs are, however, a controversial topic depending on the comparison of SEV purchase prices with different types of vehicles.

In regards to possible concerns in the interviews and the online survey, safety was mentioned as the most sensitive issue. Another stated worry in both methods (Fig. 2) was the possible switch from public transport (PT) and active modes to SEVs. Even though, this is a mentioned concern in the existing literature, there is no evidence on the potential of people switching from PT or active modes to SEVs. According to the interviewed experts, the aim of transport planning has to be on reducing the overall number of vehicles and not simply increasing it by introducing more SEVs. SEVs, however, can play a part in new mobility forms such as sharing systems. For the use, a lack of adapted infrastructure needs to be taken under consideration as currently there are no benefits for SEVs.

When talking about a sustainable mobility offer, the three pillars of social, ecological, and economic sustainability should always be considered. Although not all of them are explicitly mentioned in this paper, they often implicitly find their effect. For example, saving space for private parking would mean that the cost of parking, which is often passed on to residents, regardless of whether they own a car or not, could decrease.

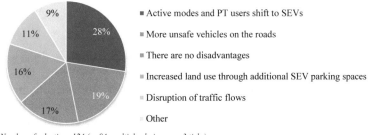

Number of selections 124 (n=84, multiple choice max. 3 ticks)

Fig. 2 Obstacles if more SEVs would operate in urban areas

4.1 Potential for Environmental Benefits

The switch from ICE vehicles to electrically propelled vehicles itself holds many benefits especially in urban areas. SEVs and BEV have a positive effect on global climate and air quality. Charging of batteries using electricity generated on renewable energies increases the positive effects and does not simply shift the CO_2 tax geographically from the city to energy production plants based on fossil fuels [13]. However, low energy consumption is still important to mitigate negative impact from renewable energy generation and due to limited available energy amount.

Contradicting climate change mitigation efforts, a trend toward larger and heavier cars can be seen since the introduction of the first serial production EVs, as the technology is advancing in terms of, e.g., higher ranges. As a negative effect, this in turn requires larger and heavier batteries using more critical raw materials for the production of the batteries and making the cars inefficient in operation. Comparing an M1 electric car, the BMW i3, with an L7e category vehicle, the Renault Twizy, the BMW i3 has a significantly higher power consumption with 13.1 kWh/100 km (measured according to VO (EU) 715/2007 [14]) than the Twizy with 8.4 kWh/100 km [15] (ADAC-Autotest). Though the consumption statements are not directly comparable due to different test cycles and should therefore not be used to quantify relative energy savings, they show the energy inefficiency of heavy vehicles bearing in mind average occupancy rates of below two persons per vehicle [16]. The inefficiency can further be illustrated by comparing the range per kWh. In this way, an SEV with one kWh can go considerably further than an electric car. Figure 3 shows the relation between the Twizy with maximum speed of 45 and 80 km/h, a Smart for two and a Mercedes-Benz B-Class EV. Adding to this, the relation between transport task and vehicle weight is highly inefficient for passenger cars based on average occupation rate. The transport task passenger transport in Germany in average means a carrying capacity of 115 kg, calculated from the average occupation rate 1.5 [16] and the average weight of an adult person 77 kg [17]. While the weight of the Twizy exceeds this transport weight 4.3 times, the B-Class exceeds the average transport task by almost 14 times.

4.2 Potentials for Land Use

Increasing population in cities intensifies the situation of scarce space and raises the question of equitable contribution of land. The current land used for transport infrastructure accounts for a large part of the total area. For example, in German large cities, transport infrastructure takes up 12% in average of the total area and even a quarter of land used for transport infrastructure compared to human settlement areas [18]. Considering the average rate of occupation, private cars are the most intense mode of mobility occupying valuable space in cities [13].

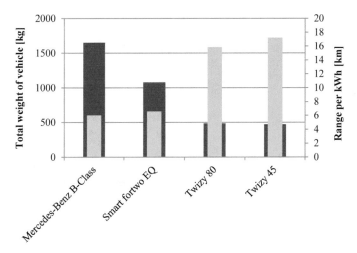

Fig. 3 Vehicle weight depending on the range in kWh, range values based on different driving cycles, for passenger cars: NEDC combined

Smaller sized vehicles take up less space on the road and require smaller parking spaces. An average parking space size of 5 m in length and approximately 2.5 m in width could be used by, e.g., three Renault Twizys or Toyota i-Roads.

Comparing space needed by SEVs, smaller M1 models, and large M1 cars, differences become apparent in parking position and in case of operation with speeds of 30 km/h (see Fig. 4). Taking into account stopping and reaction distance, the Renault Twizy needs about 20 m^2 less space than the Mercedes B-Class car in situations with 30 km/h speed. Differences in space requirement for driving are therefore minor compared to parking space potentials. Furthermore, the Mercedes B-Class enables the transport of up to five persons, resulting in a low space per person, However, with regard to average occupancy rates of 1.5 [16], this is a rather theoretical potential, used in a low percentage of trips. Even more, the figure shows the high potential for savings in land use for SEVs in parked position.

Although efforts are being made in some cities to introduce stricter regulations, e.g., in connection with the construction of new buildings, the cost of parking is mostly carried by residents. Less and smaller parking lots could decrease the overall costs for residents as well as for municipalities. Furthermore, under current circumstances SEVs spend less energy idling as they account for shorter parking search traffic than cars, because they fit in many different sized and shaped parking lots and usually can park crossways [7].

With regard to the actual potential for rededicating land used by cars, it is important to identify which user groups can switch to SEVs according to their travel behavior and how high the potential is. In the chapter "Fields of applications and transport-related potentials of small electric vehicles in Germany," a technical feasible substitution potential is calculated using data from a national household survey (MiD).

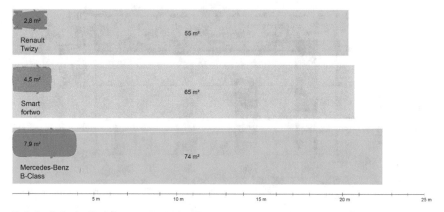

Visualization of land use in parking position and operation with 30 km/h. Road width and dimension of parking spaces are not taken into account. Areas are determined by the dimensions of the vehicles, as in most EU countries defined by law, for driving 1.5 m lateral clearance distance to single track vehicles, stopping distance and reaction distance.
Data: Renault 2020, Mercedes-Benz AG 2020

Fig. 4 Land use of different vehicle models

4.3 Safety as a Large Drawback

Although the reduced size and weight bring many benefits for the user, municipalities, and the environment, they have a higher safety risk to occupants, especially
in the event of a collision with larger vehicles. This is reflected by the results of the
quantitative survey, where concern about more unsafe vehicles on the road is the
second leading obstacle for SEVs in the opinion of the participants. In this case,
occupants of the vehicle with the lowest mass sustain the highest damage [7].
Besides disadvantages for lightweight vehicles due to physical laws that determine
accident dynamics of collisions with unequal opponents, safety features of both
lightweight and heavy vehicles influence the extent of lesions in case of a collision.

On the one hand, SEVs are not equipped with extensive safety equipment due to
the necessity of lightweight design and cost. In many countries and also according
to EU regulations, crash tests for SEVs are not required by law. Therefore, the
vehicles are equipped with minimal safety features [19]. Besides the lack of
mandatory crash tests, there are safety requirements that are laid down in EU
Regulation No. 168/2013 and the delegated EU Regulation No. 3/2014.

On the other hand, safety structures of heavy cars are not optimized for collisions
with very lightweight vehicles and usually relatively rigid. Deformation of structures that would reduce impact forces by transforming kinetic energy into deformation energy is therefore limited. This cannot be compensated by structures of
SEVs and thus leads to high deceleration of occupants in the lightweight vehicle,
causing more severe injuries. High speeds of passenger cars add to the risks in case
of an accident.

Extended safety features like airbags as standard equipment, improvement of
vehicle structures and active safety features like emergency brake assistants could

enhance safety of SEVs. Even more than technological measures, regulation could improve the situation for SEV occupants. When both SEVs and fast, heavy cars are mixed in high speed traffic, the safety risk is higher. The reduction of the maximum speed allowed, e.g., in inner-urban areas or city highways, would improve the situation. This would not only protect SEV occupants, but also vulnerable road users like mopeds, bicycles, or pedestrians. Scientific investigations show a direct link between reduction in average speed and decrease in accident numbers and crash severity, e.g., [20–23]. The extent of safety increase varies depending on initial speed and further parameters like infrastructure characteristics. For urban roads, speeding is one key factor in traffic accidents with impacts on both frequency of crashes and severity of injuries [24].

Safety issues of transport modes like bicycles and mopeds are more severe compared to SEVs; however, in contrast to SEVs, they are sold and used in large volumes. It is common consent that safety could be increased by optimized traffic regulation and infrastructural measures rather than with enhanced safety structures of these kinds of vehicles. This is similar for SEVs, even though the safety potential of vehicle technology is considerably higher and should therefore be further developed additional to regulative and infrastructural measures.

4.4   Costs of SEVs

The aspect of costs, in particular with regard to the purchase price, was discussed diversely in the qualitative analysis. In a comparison of costs, it is always very important to distinguish between the different types of vehicles. For example, the cost of owning or buying an SEV to offer in a sharing business is very high compared to e-scooters, bicycles, and some second-hand cars. Particularly in comparison with lower-priced cars, the purchase price can have a negative effect on the purchase decision, as SEVs often appear expensive with regard to limited flexibility in the transport of people and goods. However, compared to new cars, especially BEVs, they are relatively less expensive (Fig. 5).

For manufactures, the production costs for small series vehicles are significantly more expensive than mass production. However, in order to offer a vehicle to a broader user group, an attractive price is necessary. Manufacturers are therefore often faced with a dilemma. For example, by setting higher safety standards, they could offer a safer and high-quality product, but would have to set the selling price very high. For large companies developing a model for a small series vehicle in their portfolio often does not make sense as the economic risk is too high to invest in.

In the qualitative analysis, it became clear that the current situation is not favorable for SEVs in many countries, i.e., high speed limits in cities, no advantages in regards to parking or use of lanes, few incentives, few models on the market. In comparison with cars for many people, this leaves SEVs with few rewards to people considering them as relatively expensive.

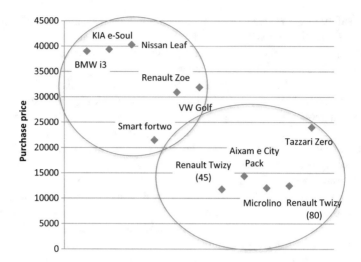

Fig. 5 Purchase prices of the top five new registrations of EV models in Germany 2018 [25], (Smart fortwo includes *EQ fortwo coupe* and *fortwo coupe ed*) and purchase price of common SEV models, Renault Twizy and ZOE purchase price calculated with battery rent for 8 years, estimated purchase price for Microlino

5 Vision of SEVs

Increasing the number of SEVs on the road, certain risks remain in the opinion of many experts. Thus, in the current traffic environment and with their lack of minimum safety requirements, they might pose some safety risks. Furthermore, a change in the mind set of how people move is needed in order to achieve that these vehicles are regarded as an equal vehicle concept for everyday mobility. Otherwise, SEVs tend to be considered at most as an additional vehicle, which makes the vehicle price appear very high. Overall, however, SEVs offer great potential for sustainable change, especially in urban areas. Scaling down weight and size of large and heavy cars has a high impact as they consume less energy and show potential to reduce space used in cities. In order to provide benefits for these vehicles to become more widely used measures including push and pull elements with the objective of replacing passenger cars with SEVs are of high importance.

As an exemplary visualization of the positive potential of SEVs for urban planning, Fig. 6 shows a vision of how urban transport could look like with SEVs. This vision is derived from statements made in the expert interviews as well as from literature research. In light of future urban landscapes especially land distribution could be modeled differently with, e.g., smaller sized parking lots, if SEVs would replace considerable numbers of passenger cars. The car would not be the dominant part which allows people-oriented city planning, creating more attractive

Fig. 6 Vision for SEVs in urban areas

surroundings with higher living standards. The introduction of SEVs into the mobility mix offers a high degree of diversification. The wider the range of mobility solutions available, the better the overall transport system can develop and harmonize with requirements of inhabitants. Therefore, SEVs can be used either as private passenger cars or within sharing schemes.

References

1. International Energy Agency: Global EV Outlook 2019: Scaling-up the transition to electric mobility. OECD (2019)
2. ACEM: Motorcycle, moped and quadricycle registrations in the European Union—2010–2018. Available https://www.acem.eu/market-data (2019). Accessed 8 March 2019
3. Hurst, D., Wheelock, C.: Neighborhood Electric Vehicles—Low-Speed Electric Vehicles for Consumer and Fleet Markets. PikeResearch, Boulder, Colorado, Research Report (2011)
4. Meuser, M., Nagel, U.: Das Experteninterview—konzeptionelle Grundlagen und methodische Anlage. In: Pickel, S., Pickel, G., Lauth, H.-J., Jahn, D. (eds.) Methoden der vergleichenden Politik- und Sozialwissenschaft, pp. 465–479. VS Verlag für Sozialwissenschaften, Wiesbaden (2009)

5. Ewert, A., Brost, M.K., Schmid, S.A.: Framework Conditions and Potential Measures for small electric vehicles on a Municipal Level. WEVJ **11**(1), 1 (2020). https://doi.org/10.3390/wevj11010001
6. Baur, N., Blasius, J. (eds.): Handbuch Methoden der empirischen Sozialforschung. Springer VS, Wiesbaden (2014)
7. Honey, E., Lee, H., Suh, I.-S.: Future urban transportation technologies for sustainability with an emphasis on growing mega cities: a strategic proposal on introducing a new micro electric vehicle statement. WTR **3**, p. 13 (2014). https://doi.org/10.7165/wtr2014.3.3.139
8. European Parliament and European Council: Regulation (EU) No 168/2013 (2013)
9. National Highway Traffic Safety Administration (NHTSA): 49 CFR 571—Federal Motor Vehicle Standards (2004)
10. Fraunhofer, I.A.O.: Standortanalyse Japan. Automobilindustrie und zukünftige Mobilitätsinnovationen. e-mobil BW GmbH—Landesagentur für neue Mobilitätslösungen und Automotive Baden-Württemberg. Available https://www.e-mobilbw.de/files/e-mobil/content/DE/Publikationen/PDF/PDF_2018/18020_Studie-Standortanalyse-Japan_RZ-Web.pdf (2018). Accessed 13 March 2018
11. Morche, D., Schmitt, F., Genuit, K., Elsen, O., Kampker, A., Friedrich, B.: Fahrzeugkonzeption für die Elektromobilität. In: Kampker, A., Vallée, D., Schnettler, A. (eds.) Elektromobilität, pp. 149–234. Springer, Berlin, Heidelberg (2013)
12. European Commission: 2007/46/EC, p. 160 (2007)
13. Santucci, M., Pieve, M., Pierini, M.: Electric L-category vehicles for smart urban mobility. Transp. Res. Part F: Traffic Psychol. Behav. **14**, 3651–3660 (2016)
14. BMW AG, The i3. https://www.bmw.de/de/neufahrzeuge/bmw-i/i3/2020/bmw-i3-ueberblick.html (2020). Accessed 25 April 2016
15. Thomas, J: Renault Twizy im Test: Ein Elektroauto macht Ernst. In: Auto Motor Und Sport, 25 July 2012. https://www.auto-motor-und-sport.de/test/renault-twizy-im-test-ein-elektroauto-macht-ernst/. Accessed 09 May 2018
16. infas Institut für angewandte Sozialwissenschaft GmbH, Mobilität in Deutschland. Kurzreport. Verkehrsaufkommen - Struktur - Trends, Bonn, Ausgabe September 2019. Available http://www.mobilitaet-in-deutschland.de/pdf/infas_Mobilitaet_in_Deutschland_2017_Kurzreport.pdf (2019). Accessed 9 Oct 2019
17. Statistisches Bundesamt, "Mikrozensus - Fragen zur Gesundheit 2017. Available https://www.destatis.de/DE/Themen/Gesellschaft-Umwelt/Gesundheit/Gesundheitszustand-Relevantes-Verhalten/Publikationen/Downloads-Gesundheitszustand/koerpermasse-5239003179004.pdf?__blob=publicationFile (2018)
18. Destatis, Bodenfläche nach Art der tatsächlichen Nutzung. Fachserie 3 Reihe 5.1 (2019)
19. Pavlovic, A., Fragassa, C.: General considerations on regulations and safety requirements for quadricycles. Int. J. Qual. Res. **9**(4), 657–674 (2015)
20. International Transport Forum: Speed and Crash Risk. Available https://www.itf-oecd.org/sites/default/files/docs/speed-crash-risk.pdf (2018). Accessed 06 May 2020
21. Nilsson, G.: Traffic Safety Dimensions and the Power Model to Describe the Effect of Speed on Safety. Lund University (2004)
22. Taylor, M., Lynam, D.A., Baruya, A.: The effect of drivers' speed on the frequency of accidents. Transport Research Laboratory, Crowthome, TRL Report TRL421 (2000)
23. Gitelman, V., Doveh, E., Bekhor, S.: The relationship between free-flow travel speeds, infrastructure characteristics and accidents, on single-carriageway roads. Presented at the World Conference on Transport Research—WTCR 2016, Shanghai. https://doi.org/10.1016/j.trpro.2017.05.398 (2017)

24. Branzi, V., Domenichini, L., La Torre, F.: Drivers' speed behaviour in real and simulated urban roads—A validation study. Transp. Res. Part F: Traffic Psychol. Behav. **49**, 1–17 (2017). https://doi.org/10.1016/j.trf.2017.06.001
25. Kraftfahrt-Bundesamt, Neuzulassungen von Kraftfahrzeugen und Kraftfahrzeuganhängern nach Herstellern und Handelsnamen, Jahr 2018 (FZ 4). https://www.kba.de/DE/Statistik/Fahrzeuge/Neuzulassungen/MarkenHersteller/n_markenHersteller_inhalt.html?nn=2601598 (2019). Accessed 12 July 2020

Courses of Action for Improving the Safety of the Powered Cycle

Luc Vinckx and Huw Davies

Abstract This paper explores the possibility to include a number of safety features from passenger cars in powered cycles with three or four wheels, whilst complying with the legal definitions and requirements, and also the legal conditions to use the bicycle lanes. The differences between technical specifications contained within EU law for pedal cycle with pedal assistance, powered cycles, quadricycles and passenger cars will be explained. Further, examples of traffic code rules with respect to the use of bicycle lanes in different countries will be discussed. Finally, the need for new safety criteria for powered cycles, replacing the existing power limit, is highlighted. In addition to the above, the need for a different technical approach to deal with the stability of 1 m wide e-bikes with a vehicle height similar to a mainstream passenger car will be discussed.

Keywords Safety · Regulation · Powered cycle

1 How to Increase Bicycle Use and at the Same Time Reduce Casualties and Injuries Resulting from Bicycle Accidents

A modal shift from passenger cars to the increased use of bicycles, e-bikes and powered cycles will help to improve air quality and at the same time reduce congestion resulting from passenger car use. The provision of a dedicated infrastructure, for example bicycle lanes, has an important role to play in this evolution. However, even with aggressive growth in infrastructure provision, it is likely that

L. Vinckx (✉)
Elephant Consult BVBA, Olmenstraat 31/1, 3080 Tervuren, Belgium
e-mail: luc.vinckx@telenet.be

H. Davies
Automotive Systems Engineering, Research Institute for Future Transport and Cities,
Coventry University, Coventry, UK
e-mail: ac2616@coventry.ac.uk

A. Ewert et al. (eds.), *Small Electric Vehicles*,
https://doi.org/10.1007/978-3-030-65843-4_2

there will be no dedicated bicycle lane for part of a given trip. On one hand, such situations lead to the so-called "black spots", where there is an increase in conflict with other road users that would result in accidents with the potential for higher injury outcome (e.g. collisions with passenger cars). On the other hand, a number of travellers, upon realising this situation and in order to avoid the risk, prefer to use their car for the entire trip. If a modal shift away from passenger cars is to be realised, then there is need to respond to these safety concerns.

At present, road safety policy—the courses of action, regulatory measures, laws, etc.—is restrictive in terms of providing appropriate solutions to the above problems. In this paper, a simple solution is proposed that will support OEMs to include a number of safety features from passenger cars in powered cycles with three or four wheels, whilst complying with the legal definitions and requirements, and also the legal conditions to use bicycle lanes. This further leverages the existing opportunity for powered cycles with three or four wheels to be driven on bicycle lanes and provides a similar safety level as other vehicle types (quadricycles, tricycles, passenger cars) allowing them to safely use existing road space used by passenger cars.

This paper is structured as follows: Sect. 2 discusses the present EU system of vehicle classification, highlighting the difference between passenger car classification and the classification of lightweight vehicles (L-category and assisted pedal cycles). Section 3 summarises the present situation regarding access to dedicate bicycle infrastructure across a number of EU member-states. Section 4 brings together vehicle classification and infrastructure access requirements to develop a new proposal that would support the goal of modal shift from passenger cars. Section 5 then concludes.

2 EU Classification of Vehicles

The EU Regulation 168/2013 [1] defines two different types of vehicles with pedals and a small electric motor. These are highlighted below:

- 'Pedal cycles with pedal assistance' which are equipped with an auxiliary electric motor having a maximum continuous rated power of less than or equal to 250 W, where the output of the motor is cut off when the cyclist stops pedalling and is otherwise progressively reduced and finally cut off before the vehicle speed reaches 25 km/h;
- 'Powered cycle (L1e-A)' as a vehicle designed to pedal, equipped with an auxiliary propulsion with the primary aim to aid pedalling with a maximum power of 1 kW and a maximum width of 1 m. The output of the auxiliary propulsion is cut off at a vehicle speed ≤ 25 km/h.

The pedal cycle with pedal assistance is excluded from vehicle type approval according to (EU) 168/2013, but it is subject to the machinery directive 2006/42/EC

[2]. Compliance with the objectives of the machinery directive can be proven by complying with a number of EN standards. The powered cycle is subject to European type approval laid down in (EU) 168/2013 and a number of "delegated acts". Further, to the above, both the 'pedal cycles with pedal assistance' and the 'powered cycle' can have two, three or four wheels.

Examples of four-wheel 'pedal cycles with pedal assistance' are the Podbike [3] (Fig. 1) and the Bio-Hybrid [4]. The Podbike is designed in accordance with EU regulations for pedal cycles with pedal assistance, and it is only slightly wider than a regular bicycle trailer. The Bio-Hybrid startup describes its vehicle as the ideal combination of a bike and a car. Essentially, it is a weather-protected four-wheel bike that can be powered by battery or pedalling. For both the Podbike and Bio-Hybrid, the electric traction motor assists the operator—as in the case of a pedelec—up to a speed of 25 km/h. As a result, both the Podbike and Bio-Hybrid are permitted to ride wherever regular bicycles are allowed.

The Podbike and the Bio-Hybrid therefore offer a unique modality. They provide the user with a small lightweight vehicle with which they can legally access bicycle infrastructure. The question is: Why has this particular modality not found the success of other modalities that have restriction on where they can be used?

Since the 'powered cycle' as well as the 'pedal cycle with pedal assistance' [1] can be built in a four-wheel version, it can be useful to compare these vehicle categories with other, more powerful vehicle categories. A comparison is made with the light quadricycle (L6e-B), the heavy quadricycle (L7e-C and L7e-A2) and the passenger car (M1). Requirements for M1 have been published by regulations 2018/858 [5]. Table 1 shows key legal parameters between five different vehicle categories for four-wheeled vehicles.

A quick review of the key legal requirements for the different vehicle categories suggests that it would be theoretically possible to create a vehicle that crosses the different categories; i.e., a vehicle could be categorised as both L6e-B and at the same time L1e-A. Indeed, it is not unknown for manufactures to develop a vehicle and to classify the vehicle in multiple categories. Examples include:

Fig. 1 Podbike (Norway) pedal cycle with pedal assistance (250 W) is announced for 2020 [3]

Table 1 Comparison of key legal parameters between different vehicle categories for four-wheeled vehicles

	Pedal cycle with pedal assistance	L1e-A	L6e-B	L7e-C/ L7e-A2	M1
		Powered cycle	Light quadricycle light quadrimobile	Heavy quadricycle heavy quadrimobile	Passenger car
EU Regulation	Excluded from (EU) 168/2013	(EU) 168/2013	(EU) 168/ 2013	(EU) 168/ 2013	(EU) 2018/858
Number of wheels	2, 3 or 4	2, 3 or 4	4	4	4
Max. length (m)		4	3	3.7/4	12
Min. length (m)		None	None	None	None
Max. width (m)		1	1.5	1.5/2	2.55
Min. width (m)		None	None	None	None
Max. mass (kg)		None	425	450[a]	None
Min. mass (kg)		None	None	None	None
Max. number of seats			2	4/2	9
Min. number of seats		(1)	1	1	1
Min. top speed (km/h)		6	6	6	25
Max. top speed (km/h)		25	45	90/None	None
Max. motor power (kW)	0.25	1	6	15	None

[a]The maximum vehicle mass in (EU) 168/2013 is based on the mass in running order without the battery for electrical propulsion. Mass in running order does not include the mass of the driver, but it includes all the liquids necessary to put the vehicle in traffic [1]

- The Renault Twizy has been on the market since 2012 and is certified as an L6e-B and L7e-C. The mass in running order is about 450 kg [6].
- The Lumeneo Smera was proposed as a small M1 with a maximum weight of 450 kg, and a vehicle width below 1 m [7].
- The Colibri from the German company Innovative Mobility Automobile (IMA), 540 kg and 1.1 m width, was presented at the Geneva motor show in 2013. The vehicle was announced to be launched in 2016, but has not yet reached the market [8].

However, classifying a vehicle into multiple categories is not that simple. If a vehicle is to be classified in different categories in additional to the requirement to comply with the physical requirements listed in Table 1 (weight, size, seats, etc.), there are additional legal requirements, mostly relating to safety or environmental performance. For classification as M1, a vehicle needs to be certified, according to UN ECE Regulation 94 on "protection of the occupants in the event of a frontal collision", UN ECE Regulation 95 on "the protection of the occupants in the event of a lateral collision" and UN ECE Regulation 127 on "pedestrian safety performance". These three regulations are considered as "defining" a vehicle platform. This means that if an existing vehicle is not compliant with the three regulations, it is almost impossible to adapt the existing platform accordingly. In this regard, it seems difficult to comply with these regulations for vehicles respecting the maximum weight for the L6e-B and L7e-C categories, especially for two-seaters in the side-by-side format, with standard side doors and a roof.

For reasons of completeness, it has to be mentioned that one-seater vehicles without a roof or standards side doors (with vertical hinges) can be certified as L6e, L7e and M1. However, even as an L-category vehicle (L6e and L7e) consideration of crashworthiness is an essential for success in the market place—for example, a number of OEMs in the L6e and L7e market space subject their product on a voluntary basis to crash testing to provide confidence to the consumer [9].

Many other regulations apply on a mandatory basis to the different vehicle categories mentioned in this article, but they do not impact the basic mechanical structure of the vehicle platform. Of course, they might add weight and therefore adding or mandating these technologies might lead to a non-compliance of the maximum weight for L6e-B and L7e-C.

The requirement to consider crashworthiness has a penalty in terms of vehicle mass. Energy absorbing structures, safety equipment and the need to design a vehicle to include crush space all increase vehicle mass. Further, the inclusion of crashworthiness considerations in light and heavy quadricycles leads to a requirement for higher motor power in order to overcome the increase in mass.

The Renault Twizy, Lumeneo Smera and Colibri represent market acceptable solutions in the light and heavy quadricycle categories. The Podbike and Schaeffler Bio-Hybrid represent solutions in the 'pedal cycles with pedal assistance' category. When looking at these five vehicles or prototypes, belonging to two different "families", it appears that a common bodywork, with a width not more than 1 m could be used in both groups. For the L1e-A, the overall maximum width is 1 m, for L6e-B and L7e-C the maximum width is 1.5 m and for M1 the maximum width is 2.55 m.

From the argumentation above, it becomes clear that a car based on an e-bike powertrain is possible.

The concern is that three- or four-wheel e-bikes are not yet being promoted as vehicles allowed to be used on bicycle lanes. This could be seen as a new modality: They can use bicycle lanes wherever present, but are forced to use the car lanes when there is no bicycle lane. The latter is a concern to users and can be a reason for not using a bike, but a car. A solution would be to include the safety elements of the light and heavy quadricycles in an L1e-A vehicle. It is possible to develop a family of vehicles, using the same bodywork with a maximum width of one metre. Different variants of this vehicle could be certified in different vehicle classes.

Different track widths could be used for L1e-A, L6e-B/L7e-C and M versions, in line with the different maximum widths for these categories.

Assuming that an L1e-A vehicle could be developed sharing many of the safety elements of vehicles certified in the L6e-B, L7e-C category or M1 class, it is important:

- To investigate whether that vehicle (the improved L1e-A category vehicle) would still be allowed to be driven on a bicycle lane (addressed in Sect. 3)
- To determine the maximum weight possible for an L1e-A variant and hence if the extent to which the safety of the L6e/L7e/M1 can be kept (addressed in Sect. 4).

3 Use of the Bicycle Lane: Differences Between Countries

Within the EU, technical regulations have been harmonized. For the traffic rules, there has not been the same level of harmonisation. The only elements of the traffic rules harmonised at EU level, known to the authors, are the driving licence and minimum age, defined in EU Directive 2006/126/EC [10].

When it comes to the rules with respect to the use of the bicycle lane, some differences can be detected between member-states. Examples are:

- Germany: L1e-A are not allowed the standard bicycle lanes. But there are some "special" bicycle lanes where they are allowed ("E-Bikes allowed" or "Mofas allowed") [11]
- Belgium: L1eA are allowed on all bicycle lanes [12]
- Netherlands: Electric bikes with top speed up to 25 km/h are allowed to be driven on bicycle lanes; therefore, this includes L1eA [13]
- UK: With more than 250 W auxiliary power, a bicycle is not allowed on bicycle lanes [14].

Further, markets outside of the EU that still share the same vehicle classification requirements can be considered:

- Norway (not part of EU): All bicycles, e-bikes with power assistance up to 250 W can drive on the bicycle lanes [15].

The observation from the above is that 'pedal cycles with pedal assistance' and 'powered cycles' are treated differently depending on the market. This causes significant complications in the promotion and development of new modalities.

As a first step, harmonisation of traffic rules, e.g., concerning the use of bicycle lanes, would be helpful for market development of new vehicle concepts, e.g., based on the L1e-A regulatory framework. Second, the adaption of traffic rules to facilitate and promote the adoption of new modalities that support the move to cleaner and more sustainability mobility would be beneficial. At this point, the question arises how these requirements/rules should look like.

4 Calculation of the Maximum Mass of a Powered Cycle

In order to estimate the total vehicle weight that still allows an L1e-A vehicle to be driven at a speed of 25 km/h with an electric motor power of 1 kW, we made some basic calculation of the instantaneous power as a function of the aerodynamic resistance, vehicle dimensions, total vehicle weight, road friction, vehicle speed and the slope of the street.

The formula to calculate the instantaneous power to propel a road vehicle can be derived by combining the definition of mechanical power and the specific formula's for aerodynamic force F_a, the rolling resistance force F_r and the component of vehicle weight alongside the slope of the road, F_w.

P = Instantaneous power at the wheels
v = Instantaneous vehicle speed
Definition of power: P = (sum of the forces exerted on the vehicle), v

$$P = (F_a + F_r + F_w).v \tag{1}$$

When a vehicle is driving on a slope with constant speed (no acceleration), three forces are active in the direction of motion of the vehicle:

– The aerodynamic resistance force F_a is proportional to the frontal surface area A, to the square of the vehicle speed, to the aerodynamic coefficient (C_x), to the air density ρ (normally 1.2 kg/dm^3) and to the square of the vehicle speed:

$$F_a = \frac{1}{2}.\rho.C_x.A.v^2 \tag{2}$$

– The rolling resistance force F_r is proportional to the vehicle mass m, the gravitational acceleration g and the rolling resistance coefficient μ. A normal value for μ on wet road surfaces is 0.015.

$$F_r = \mu.m.g \tag{3}$$

– The component of the vehicle weight is parallel to the road surface, F_w For small values (e.g., 0.05) of the slope θ: $tg(\theta) = \sin(\theta) = \theta$. So, the formula can be simplified into:

$$F_w = m.g.\theta \tag{4}$$

Combining the expressions for the three forces (2), (3) and (4) in formula (1) for the power, we find the following expression:

$$P = (F_a + F_r + F_w) = F_a = \frac{1}{2}.\rho.C_x.A.v^3 + \mu.m.g.v + \theta.m.g.v \qquad (5)$$

Another way to write the formula could be:

$$P = P_a + P_r + P_w \qquad (6)$$

This formula reads as follows: The instantaneous mechanical power is the sum of:

- The power necessary to compensate the aerodynamic resistance,
- plus the power necessary to compensate the rolling resistance,
- plus the power to compensate the weight component in case of a slope.

The final power output of the electric motor needs to take into account the parasitic losses in the powertrain. With p the parasitic losses and P_{em} the power of the electric motor, our aim is:

$$P < (1 - p).P_{em} \qquad (7)$$

This formula can be easily implemented in a spreadsheet. In Table 2, one particular set of input values is proposed. The calculation is done with a total mass of 500 kg, assuming that with this mass, sufficient safety technology can be integrated to guarantee a "safety level" similar to the quadricycles and small cars. The corresponding output results are given in Table 3. The output shows that the "heavy" version of the powered cycle L1e-A is able to drive at 25 km/h on a slope of 5% with a total mass of 550 kg.

The conclusion from the calculation: If a narrow vehicle (vehicle width <1 m) can be certified as an L6e, L7e or M1, the bodywork can be used as the basis for an L1e-A vehicle.

Table 2 Some reasonable estimates to calculate the power need for an L1eA at top speed

Parameter	Estimate	Units
Vehicle speed v	7	m/s
	25	km/h
Frontal surface A	1.5	m^2
Aerodynamic coefficient C_x	0.4	–
Total mass (vehicle, passenger, luggage) m	550	kg
Rolling resistance μ	0.015	–
Gravitational acceleration g	9.81	m/s^2
Slope θ	0.05	(rad)
Parasitic losses p	0.15	–

Table 3 Output of the calculation (based on the input from Table 2)

Power to compensate the aerodynamic resistance	123 W
Power to compensate the friction	567 W
Power to compensate the slope	270 W
Total mechanical power needed (P)	960 W
Legal maximum power for electric motor L1e-A (EU)	1000 W
+Human power, modest estimate	200 W
Total power available	1200 W
Total power available after parasitic losses $((1-p).P_{em})$	1020 W

5 Conclusion

Small electric vehicles (SEVs) provide an opportunity to respond to environment and mobility concerns. The limitation is safety. Dedicated infrastructure, for example bicycle lanes, support the market acceptance of SEVs by providing an environment that reduces risk to the user by removing conflict with larger, more aggressive, collision partners such as passenger vehicles.

However, for end-to-end journeys, an SEV user will be required to use a mix of dedicated and shared infrastructure. This creates a conflict. When SEVs are used in a mixed infrastructure, consumers require safety features that increase mass and hence power. If they are to be allowed access to the dedicated infrastructure, the vehicle power and hence mass must be limited.

In Chap. 3, it was shown that some countries allow the powered cycle on a bicycle lane, but in many countries, the right to be driven on the bicycle lane is reserved to power-assisted e-bikes with a maximum power of 250 W. At this juncture, it is uncertain what is the origin and the meaning of this 250 W limitation. Furthermore, this limitation on power removes an opportunity for creating cost-effective mobility solutions. In Chap. 4, it was shown that there is an interesting opportunity to develop a variant of the powered cycle that can share a lot of components with quadricycles and cars. Because of the positive effect of economies of scale, this could be a route to improve the economics of SEVs.

The authors therefore question whether it is appropriate to exclude narrow vehicles (<1 m) from the bicycle lanes irrespective of maximum power providing that they that do not drive faster than 25 km/h.

In short, we conclude and propose:

- Component and/or platform sharing between L1e-A, L6e, L7e and M1 is possible
- International harmonisation of rules with respect to the use of bicycle lanes for narrow vehicles restricted at 25 km/h, but with power >250 W, furthermore, approx. 1 or 2 kW, should be considered in this regard.

- The power limit of 250 W as a safety criterion should be replaced by a different criterion. More research is needed to identify the need and eventually to develop such a criterion.

References

1. Regulation EU No 168/2013 of the European Parliament and of the Council of 15 January 2013 on the approval and market surveillance of two- or three-wheel vehicles and quadricycle. Official Journal of the European Union, L60
2. Machinery Directive. Directive 2006/42/EC of European Parliament and of the Council of 17 May 2006 on machinery, and amending Directive 95/16/EC (recast). Official Journal of the European Union, L157
3. Forus Lab: Pre-Order Podbike for Delivery in Europe, https://www.podbike.com/en/pre-order-podbike-delivery-in-eu-and-eea/ (2020). Last accessed 1 Sept 2020
4. Schaeffler Bio-Hybrid GmbH: Schaeffler Bio-Hybrid GmbH with world premiere at CES 2019. https://www.schaeffler.com/remotedocuments/en_2/_global/download_pdf_rtf/pdf/pressrelease_85726913.pdf (2018). Last accessed 1 Sept 2020
5. Regulation EU No 2018/858 of the European Parliament and of the Council of 30 May 2018 on the approval and market surveillance of motor vehicles and their trailers and of systems, components and separate technical units intended for such vehicles, amending Regulations (EC) No 715/2007 and (EC) No 595/2009 and repealing Directive 2007/46/EC
6. Renault: https://www.renault.fr/vehicules-electriques/twizy.html. Last accessed 1 Sept 2020
7. Lumeneo Smera: https://www.avem.fr/actualite-la-lumeneo-smera-au-mondial-de-paris-2010-1760.html. Last accessed 1 Sept 2020
8. Colibri, I.M.A.: https://ecosummit.net/uploads/eco13_151013_1620_thomasdelossantos_innovativemobility.pdf. Last accessed 1 Sept 2020
9. Davies, H., Turrell, O.: Ped-elec: development of a new value chain approach to the provision of an urban mobility solution. Paper presented at 32nd Electric Vehicle Symposium (EVS32) Lyon, France, May 19–22 (2019)
10. Directive 2006/126/EC of the European Parliament and of the Council of 20 december 2006 on driving licences (Recast). Official Journal of the European Union, L403
11. § 39 Verkehrszeichen (Straßenverkehrs-Ordnung) v. 06.03.2013, BGBl. I, p. 367, https://dejure.org/gesetze/StVO/39.htm. Last accessed 1 Sept 2020
12. FOD Mobiliteit en Vervoer: FAQ elektrische fietsen. https://mobilit.belgium.be/nl/wegverkeer/wetgeving_en_reglementering/faq_elektrische_fietsen (2015). Last accessed 1 Sept 2020
13. Visser A.: Onbekende categorie e-bikes met meer power. https://www.tweewieler.nl/elektrische-fietsen/nieuws/2018/02/elektrische-fietsen-met-hoog-motorvermogen-bezig-aan-opmars-10133805?_ga=2.17703609.1060690443.1598957087-1867613040.1598957087 (2018). Last accessed 1 Sept 2020
14. UK Government: Electric bikes: licensing, tax and insurance. https://www.gov.uk/electric-bike-rules (2020). Last accessed 1 Sept 2020
15. Forskrift om krav til sykkel, 01.04.1990, https://lovdata.no/dokument/SF/forskrift/1990-02-19-119. Last accessed 1 Sept 2020

Velomobiles and Urban Mobility: Opportunities and Challenges

Geoffrey Rose and Alex Liang

Abstract As the transport challenges facing urban areas intensify, innovative solutions are required to address the social, economic and environmental impacts arising from overreliance on private motor vehicles. Velomobiles offer a range of advantages but do not feature on the radar screen of urban transport policy makers. This chapter explores the challenges and opportunities of increased adoption of velomobiles as an urban mobility option. A scan of existing velomobiles is used to define typical characteristics of these vehicles and place them into perspective against relevant travel options before they are assessed in the context of typical vehicle regulations and facility design guidelines. The opportunities and challenges associated with greater adoption of velomobiles in the context of urban travel are examined through the lenses of technology adoption and the sociotechnical framing of independent travel options. Shared mobility is identified as one potential way of broadening the base for velomobile adoption in urban areas.

Keywords HPV/human-powered vehicles · Velomobiles · Vehicle classification · Infrastructure design guidelines · Shared mobility

1 Introduction

As urban populations are growing in many cities, the pressure on urban transport systems is intensifying. Those pressures are compounded by the reliance on car-based mobility—the prevailing transport monoculture in many cities [1]. The emphasis on in-active transport, and the inevitable traffic congestion and emissions, results in lower productivity and considerable negative external costs that are not directly born by users [2, 3]. Responses to these challenges within the transport industry have presented opportunities for disruptive technologies. Disruptive technologies influence and add value to established markets, often displacing older

G. Rose (✉) · A. Liang
Monash Institute of Transport Studies, Monash University, Melbourne 3800, Australia
e-mail: Geoff.Rose@monash.edu

© The Author(s) 2021
A. Ewert et al. (eds.), *Small Electric Vehicles*,
https://doi.org/10.1007/978-3-030-65843-4_3

29

technologies or methods. In the transport sector, such technologies may influence the mode choice of individuals which will have implications for the social, economic and environmental impacts arising from the transport system.

Many light electric vehicles are examples of potentially disruptive technologies [4] in the context of urban transport where these technologies have the potential to disrupt the dominant role of private automobile ownership and use in urban mobility. Examples of disruptive transport technologies include shared mobility such as car sharing and bike sharing [5] as well as emerging motorised personal mobility devices (PMD) such as electric-power-assisted bicycles [6], electric scooters, Segways and other self-balancing motorised devices. Laws and regulations have a large impact on the adoption and usage of various disruptive technology options. The focus of this chapter is on velomobiles, which are one form of PMDs.

A Web search of the images of a 'velomobile' will bring up a range of images of two- or three-wheeled vehicles with their shape suggesting consideration of aerodynamics. One example is shown in Fig. 1.

A definitive definition of a velomobile is problematic and as noted by van de Walle [7, p. 69]:

> even the people closely involved with velomobiles cannot accurately define a velomobile and there are plenty of discussions on what exactly constitutes a velomobile. Although a velomobile is very different from a bicycle, enthusiasts usually describe a velomobile in relation to the (recumbent) bicycle.

Non-enthusiasts who see a velomobile and are not sure at first what it is and might reach the conclusion that it is essentially:

> a special bicycle .... an expensive, heavy, complex, large and difficult to park bicycle with extra wheel(s) and a body on top of it. van de Walle [7, p. 81]

The platform Wikipedia [8] and [7] offer the following definition which picks up key attributes and characteristics:

> A velomobile, is a human-powered vehicle (HPV) enclosed for aerodynamic advantage and/or protection from weather and collisions. They are similar to recumbent bicycles and tricycles, but with a full fairing (aerodynamic or weather protective shell).

Fig. 1 Side and front view of a velomobile (own image)

While there is no doubt that velomobiles are human powered, there is growing interest in electric-assisted velomobiles. Other authors note the relevance of luggage-carrying capacity [7] which does not feature in the Wikipedia definition. The issue of safety is considered later, but it is not clear that the design of velomobiles, except perhaps for those aspiring to world record speed titles where rider safety is an explicit consideration, inherently provides any substantial amount of collision protection. The shell/body is provided primarily for aerodynamics and weather protection rather than structural integrity in the case of a collision.

Cox [9] noted that "Within the existing framework of transport options, the velomobile has a heavily circumscribed market as a symbol of the social elitism amongst cyclists" and added that "If the velomobile is itself a marginalised form of cycle, then it is difficult to envisage a greater future role than its current limited market".

While a decade and a half have passed since those words were written, they still hold true today. Yet velomobiles offer obvious opportunities from the perspective of enhancing the sustainability of our urban transport systems. They have low energy requirements, have passenger-carrying capacities that meet the needs of most trips in urban areas, have a smaller spatial footprint than a conventional motor vehicle and provide exercise and associated health benefits for the user.

The aim of this chapter is to explore the challenges and opportunities of increased velomobile usage in the context of urban travel. The approach is broken into three distinct components. First, an international scan of existing velomobiles is used to identify typical characteristics of these vehicles and place them into perspective against relevant travel options. Second, velomobiles are assessed in the context of typical vehicle regulations and facility design guidelines and finally the opportunities and challenges associated with greater adoption of velomobiles in the context of urban travel are examined. A short conclusions section wraps up the chapter.

2   Velomobile Characteristics

Velomobiles have varying attributes and their characteristics are ultimately dependent on the design and function of the vehicle. Velomobiles may be designed as an alternative mode of transport, for the purpose of attaining maximum speeds or as a recreational vehicle. An appropriate starting point is to appreciate the typical characteristics of velomobiles, including their physical geometry or spatial footprint as well as the speed profiles and operating characteristics of these vehicles. Table 1 summarises indicative values for a range of parameters. The focus here is on the general trend across different types of vehicles rather than focusing on the measurement accuracy of one particular parameter for a specific vehicle.

In general, velomobiles have a greater spatial footprint compared to road bikes, e-bikes and recumbent bicycles as given in Table 1. The dimensions of both human-powered and electric-assisted velomobiles vary greatly as velomobiles are

Table 1 Typical characteristics of various mobility devices (*HP* = Human-powered, *EA* = Electric assist. *Source* Manufacturer's web sites)

	Road bike	E-bike	Recumbent trike**	HP velomobile	EA velomobile
Width (m)	0.61	0.61	0.83–0.88	0.56–1.02	0.76–1.5
Length (m)	1.5–1.8	1.5–1.8	1.71–2.08	1.98–2.85	2.17–3
Turn radius	Low	Low	Med 1.45–3 m	Med-high 1.5–6.5 m	Med-high 1.5–6.5 m
Weight (kg)	6.8–17	23	14.5–21.8	15–80	32.5–120
Eng. power*	N/A	<250 W	N/A	N/A	<250 W

*Engine power limit under Australian and European regulations. Other countries may vary due to different regulations
**Catrike and Terratrike recumbent trikes

designed for different functions and design requirements [10]. Smaller and lighter velomobiles are generally used for racing purposes whilst mid-size and large-size velomobiles are more suited as a mode of transport [11]. Some designs incorporate two seats or additional room for storage space and therefore have larger spatial footprints.

There are velomobiles that can travel at very high speeds. The world record of a human-powered velomobile is 144.17 km/h [12]. The high speeds are due to a low centre of gravity and an aerodynamic shell of the vehicle [13]. Table 2 shows a comparison of speeds between bicycles and velomobiles. An average healthy adult can deliver 100 W of power on a bicycle and maintain that for approximately one hour. By contrast, 250 W is the power output of a well-trained cyclist. It is evident that velomobiles can travel at higher speeds with the same amount of energy input

Table 2 Speed comparison between typical bicycles and velomobiles (*Std* = standard, *BP* = best practice)

	Poorly maintained bike*	Good regular bike**	Std velomobile***	Racing bike****	BP velomobile*****
Flat road, 250 W	23.5 km/h	29 km/h	41 km/h	37.5 km/h	50 km/h
Flat road, 100 W	15 km/h	20.5 km/h	28 km/h	27 km/h	34 km/h
5% uphill, 150 W	6.5 km/h	9.7 km/h	8.6 km/h	11.6 km/h	9 km/h
Power require to ride 30 km/h	444 W	271 W	115 W	137 W	79 W

*Typical bicycle used for short distance transportation with rusty chain, underinflated tires, bad riding position, no gearing. Rough indication only
**Bicycles for transport, including fenders luggage, upright rider position
***Flevobike Alleweder
****UCI compliant bike, deep racing posture, cycling clothes
*****Velomobile.nl

Table 3 Kinetic energy generated for bikes and velomobiles

	Good regular bike*	Racing bike**	Standard velomobile***	Electric-assisted velomobile****
Flat road, 250 W	5004 J	4781 J	39,924 J	95,313 J
Flat road, 100 W	2500 J	2479 J	18,620 J	44,073 J

*Average weight of bicycle from Table 1, speeds from Table 2
**Lowest weight of bicycle from Table 1, speeds from Table 2
***Average weight from Table 1, speeds from Table 2
****Average weight from Table 1, speeds assumed to be of best practice velomobiles from Table 2

compared to bicycles, with the exception in uphill situations. Electric-assisted velomobiles are especially useful on steep hills.

Kinetic energy management is recognised as a critical factor in the context of road safety [14] since it is associated with the potential for injury in the event of a crash. Using the data from Tables 1 and 2, the kinetic energy generated for velomobiles and bicycles has been estimated and is given in Table 3. Kinetic energy is calculated as mass times velocity (speed) squared. Regular bicycles and racing bikes generate similar amounts of kinetic energy as the former are heavier and slower and the latter are faster but lighter. Standard velomobiles produce considerably higher amounts of kinetic energy compared to bicycles with the same amount of energy input due to higher speeds and increased weight. The difference in kinetic energy is even greater for electric-assisted velomobiles due to the same reasons.

3   Opportunities and Challenges

As noted in the introduction, urban transportation systems currently face substantial challenges. In that context, velomobiles are a potentially attractive option to enhance the sustainability of urban transport systems. They are space and energy efficient in operation and can be emission free (if batteries for power assistance are charged from renewable energy). From a sustainability perspective, there is an added benefit in terms of resource consumption. Velomobiles consume far less resources in manufacture than a motor vehicle given that 27 velomobiles are comparable in mass to one small car [7, p. 74]. The ability to travel at higher speeds without great effort, which is even more evident for electric-assisted velomobiles, improves comfort and efficiency when compared to bicycles and recumbents. The weather-resistant body shell of velomobiles also adds comfort for the rider. Some variants of velomobiles are also capable of storing luggage which adds utility benefits, whilst others can also carry a passenger. These properties reflect some of the benefits that private motor vehicles provide, whilst still delivering the environmental and health benefits of a bicycle. Despite these many benefits, velomobiles remain a marginalised mode, with demand low and small-scale manufacture resulting in high production costs.

Innovative approaches to design and manufacture may present opportunities to lower production costs. This could include 'growing' body components for velomobiles from bamboo [15] or a combination of modular design, open sourcing, material recycling or additive manufacture [16]. However, there is an inevitable link between costs and demand. Production costs will not fall through economies of scale till production volumes increase. Yet that will not happen till demand increases but that demand is suppressed by high prices.

To date, the velomobile has engaged technology enthusiasts and visionaries, groups which characterise the early adopters of technology [17]. The challenging step in the context of technology adoption is going beyond those early adopters to engage the pragmatists who make up the early majority and that leap has been characterised as 'crossing the chasm' and calls for flexible and innovative product development and manufacturing alongside marketing to achieve more substantial product-market fit [17].

As yet there is little sign that velomobiles are building anywhere near the sort of momentum required to cross the chasm and achieve broader market appeal. Van de Walle's (2014) seminal work on the interrelatedness of social and technical aspects of velomobiles, a concept that he articulates as the sociotechnical frame of this technology remains relevant in that context today. Van de Walle conceptualised a change in the evolinear social frame which traditionally sees a linear progression from the bicycle to the assisted bicycle to the motorcycle to the motor car. His matrix frame seeks to position the bicycle to velomobile mobility transition as comparable to the motorcycle to motor car transition. In the context of contemporary thinking about urban mobility, those pairs reflect low and high negative external costs or externalities (Fig. 2). Externalities arise due to, e.g., congestion or emissions and are costs that impact on third parties but are not reflected in the prices paid by users [18].

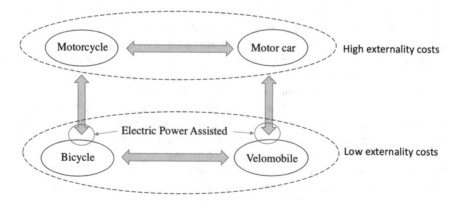

Fig. 2 Matrix of personal mobility options. *Source* Modified from [7]

One technology, which has made progress in crossing the chasm, is the electric-power-assisted bicycle even though van de Walle's assessment in 2004 was not optimistic:

> practice tells us that the users of 'assisted bicycles' remain marginal actors to the bicycle sociotechnical frame. The assisted bicycle is in a similar process as the recumbent bicycle to become accepted as legitimate variant of a new bicycle sociotechnical frame, modified from the old established one that excluded the assisted bicycle. [7, pp. 72]

The electric-power-assisted bicycle, or e-bike, has grown substantially in market share in the last decade and a half [19] and is being viewed positively for the opportunities that it present to enhance urban transport options [6]. Concerns have been expressed about the impact on e-bikes on levels of physical activity. However, the results of recent research are very positive indicating substantial increases in physical activity for users who switch from a car and limited net losses from those switching from cycling because of increases in overall travel distance [20]. Whilst in some respects, the e-bike highlights that opportunities remain from mobility options to gather traction, and there are other factors which are likely to act as barriers to growth in velomobile adoption and use.

Velomobiles typically meet the regulations associated with bicycles or Pedalelecs [21]. Consequently they face lower regulatory barriers than if their characteristics meant they were reclassified as quadricycles, mopeds or motor vehicles since they would therefore need to meet tighter design regulations. In Australia, as in Europe, they are classified as bicycles so long as the auxiliary power is less than 250 watts, and the maximum power-assisted speed is restricted to 25 kph. However, velomobile riders would be required to wear a bicycle helmet in Australia which is one jurisdiction that has mandatory helmet legislation. Even though some velomobiles can travel at high speeds, they would be forbidden from operating on urban freeways in many countries including Australia.

As velomobiles are classified as bicycles, the facility design guidelines that apply to bicycles need careful consideration. Velomobiles are generally wider than bicycles but can still fit in standard bicycle lanes and shared-use paths, although the manoeuvring space and lateral clearance is lower [22]. Cycling infrastructure is typically designed for bicycles which have a smaller spatial footprint compared to velomobiles. Although it is legal to operate velomobiles in the same locations as bicycles, it may be difficult to move efficiently or safely due to their physical characteristics and the existing infrastructure. A typical velomobile will fit in a bicycle lane but the standard lateral clearance that is used for manoeuvring, which is provided for regular bicycles, may not be adequate for velomobiles. Shared-use paths allow individuals to travel in both directions which may be a concern because of the increased width of velomobiles. Existing infrastructure may not be adequate to cater for velomobiles overtaking other users of the shared-use path or the con-current use of a mixture velomobiles and bicycles.

Safety is also an important consideration since collisions may have serious consequences. Velomobiles are capable of travelling at higher speeds than bicycles and with significantly less effort as given in Table 2. Hence, one potential barrier for velomobiles is likely to be real or perceived issues with safety. Little is known about the safety performance of velomobiles although examination of single vehicle velomobile crashes in Germany highlights speed as a contributing factor [23].

The weight of velomobiles is generally greater than bicycles and recumbents, particularly electric-assisted velomobiles. Speed and weight both influence the amount of kinetic energy which would need to be dissipated in the event of a crash. Managing the dissipation of kinetic energy in that case is critical in determining the risk of serious or fatal injuries [14, 24]. Crash rates are not solely a function of higher speed but rather increases in variances of speed [25], and hence, there could be greater concerns where there is a mix of users such as pedestrians, cyclists and velomobile riders. The braking regulations for bicycles also apply to velomobiles but may not be suitable due to the greater speeds of velomobiles. This raises potential safety issues as velomobiles may need brakes to be of higher capabilities to ensure that the rider can stop in the event of an incident.

Since the existing infrastructure is geared towards bicycles, bicycle paths and shared-use paths which may be under designed for the speed and braking capabilities of velomobiles. Bicycles are able to make sharper turns as they have a smaller turn radius and travel at lower speeds. Velomobiles may require larger curve radii, and additional sight distance may need to be provided along shared-use paths to create a safe environment for all users. Of course, this is dependent on where velomobiles are ridden and travel surveys do not currently provide insight in that context because of the low incidence of velomobile use in the population. Their classification as bicycles means they can legally be ridden on bicycle facilities. However, incompatibilities can arise between their performance characteristics (e.g., maximum speeds) and the design characteristics of those facilities, for example, in relation to the horizontal curve radii which are designed on the expectation of lower maximum speeds.

One other area which may present a barrier to greater adoption relates to parking. While classified as bicycles, the larger spatial footprint of velomobiles makes them incompatible with common bicycle parking infrastructure. This is partially illustrated by the example provided in Fig. 1 where the capacity of the bicycle parking hoops is reduced by the parked velomobile. The suspension systems of velomobiles can be designed to facilitate vertical parking and storage [13] and that may present other opportunities to overcome parking challenges.

The infrastructure issues raised above are potential barriers to use, but the challenge of stimulating demand remains. Cox [9] highlighted the need to explore the possibilities for sociable velomobiles to encourage adoption. Reflecting emergence of the shared economy, some commentators see shared mobility, such as car sharing and bike sharing [5], as one of the three pillars to underpin the transition to a sustainable transport system along with electrification and automation [1]. In that context, the emergence of shared mobility options based on a velomobile may assist in helping this innovative form of urban mobility to gain momentum. One example is Veemo (Velometro Mobility), a one-way sharing network of three-wheeled, electric-assisted velomobiles being developed in Vancouver [26]. Veemo brings together a new vehicle with a shared vehicle mobility platform to offer a new option in the context of shared mobility. Systems such as that may help to 'normalise' velomobiles within the context of urban mobility. This is analogous to what systems of shared electric scooters have done in cities around the world to normalise this type of mobility option which has helped to stimulate private ownership of e-scooters where users perceive that to be a regular part of their urban mobility system. The emergence of shared velomobile systems may help to advance the normalisation of velomobile technology in the context of urban travel.

4 Conclusion

Velomobiles have the potential to provide a mode choice alternative in urban environments since they provide some of the benefits of both bicycles and motor vehicles. Human-powered and electric-assisted velomobiles have varying physical and operational characteristics. In comparison with bicycles, they have a larger spatial footprint, can travel faster with less energy and are heavier. They have the advantage of still being classified as bicycles (depending on their characteristics). Therefore, the same regulations and laws that apply to regular bicycles currently apply to velomobiles. There are many implications for velomobiles regarding existing infrastructure and facility design guidelines as these are designed to cater for the performance envelope of regular bicycles. High speeds and a larger spatial footprint, combined with the lack of appropriate infrastructure, raise safety issues both on road and on shared-use paths. The challenge as new velomobile entrepreneurs emerge is determining whether the community and the transport profession will continue to regard them in the same category as bicycles particularly when the spatial footprint makes them incompatible with many existing bicycle facilities. Current regulatory frameworks which are based on vehicle descriptions rather than performance-based standard present a risk for disruptive technologies like velomobiles. The short-term risk for even current velomobile developers is that regulatory responses to other innovative modes may have unintended consequences for the use of velomobiles by placing new restrictions on where and when current, but less common vehicles, are permitted to operate. The emergence of velomobiles as the basis for systems of shared mobility may be a valuable stimulus for velomobile adoption. Shared-use systems could play a part in helping velomobile technology to cross the chasm to wider adoption and emerge as a more mainstream urban travel option.

References

1. Sperling, D., Gordon, D.: Two Billion Cars: Driving Towards Sustainability, p. 304. Oxford University Press, New York (2009)
2. Barth, M., Boriboonsomsin, K.: Real-world carbon dioxide impacts of traffic congestion. Transp. Res. Record: J. Transp. Res. Board **2058**(1), 163–171 (2008)
3. Weisbrod, G., Vary, D., Treyz, G.: Measuring economic costs of urban traffic congestion to business. Transp. Res. Record: J. Transp. Res. Board **1839**, 98–106 (2003)
4. Christensen, C. M.: The innovator's dilemma: when new technologies cause great firms to fail. Harvard Business School Press, Boston, Mass (1997)
5. Shaheen, S., Cohen, A., Chan, N., Bansal, A.: Chapter 13—sharing strategies: carsharing, shared micromobility (bikesharing and scooter sharing), transportation network companies, microtransit, and other innovative mobility modes. In: Transportation, Land Use, and Environmental Planning, pp. 237–262. E. Deakin, Elsevier (2020)
6. Rose, G.: E-bikes and urban transportation: emerging issues and unresolved questions. Transportation **39**, 81–96 (2012)
7. van De Walle, F.: The Velomobile as a vehicle for more sustainable transportation. In: Royal Institute of Technology Department for infrastructure (2004)
8. Wikipedia: Velomobile. https://en.wikipedia.org/wiki/Velomobile (2020). Accessed 26 April 2020
9. Cox, P.: Framing consumption: social impediments to velomobile adoption. In: Proceedings of the 5th European Velomobile Seminar: Towards Commercial Velomobiles (2004)
10. Pehan, S., Kegl, B.: Efficient Velomobile Design. Appl. Mech. Mater. **806**, 232–239 (2015)
11. Hiles, D.: 30 Iconic velomobile designs from the past 85 years. https://www.icebike.org/30-iconic-velomobile-designs-from-the-past-85-years/ (2015). Accessed 4 Sept 2018
12. International Human Powered Vehicle Association: Official speed records. http://www.ihpva.org/hpvarech.htm (2020). Accessed 26 April 2020
13. Sørensen, P.H.: VELOMOBILE: Redefined. University of Stavanger, Norway (2014)
14. Corben, B., Cameron, M., Senserrick, T., Rechnitzer, G.: Development of the visionary research model—application to the car/pedestrian conflict. Monash University Accident Research Centre, Report No. 229 (2004)
15. Vittouris, A., Richardson M.: Designing vehicles for natural production: growing a velomobile from bamboo. In: Proceedings, Australasian Transport Research Forum (2011)
16. Richardson, M.: Fab Velo. PhD in Industrial Design. Monash University (2013)
17. Moore, G.A.: Crossing the Chasm, pp 288. HarperCollins (2014)
18. Cowie, J.: The economics of transport—A theoretical and applied perspective, pp 384. Routledge (2010)
19. Jones, T., Harms, L., Heinen, E.: Motives, perceptions and experiences of electric bicycle owners and implications for health, wellbeing and mobility. J. Transp. Geogr. **53**, 41–49 (2016)
20. Castro, A., Gaupp-Berghausen, M., Dons, E., Standaert, A., Laeremans, M., Clark, A., Anaya-Boig, E., Cole-Hunter, T., Avila-Palencia, I., Rojas-Rueda, D., Nieuwenhuijsen, M., Gerike, R., Panis, L. I., de Nazelle, A., Brand, C., Raser, E., Kahlmeier S., Götschi, T.: Physical activity of electric bicycle users compared to conventional bicycle users and non-cyclists: insights based on health and transport data from an online survey in seven European cities. In: Transportation Research Interdisciplinary Perspectives 1 (2019)
21. Lieswyn, J., Fowler, M., Koorey, G., Wilke, A., Crimp, S.: Regulations and safety for electric bikes and other low-powered vehicles. NZ Transport Agency Research Report **621**, 1–182 (2017)
22. Austroads Guide to Road Design: Part 6A: Pedestrian and Cyclist Paths, Austroads (2017)
23. Bunte, H., Hipp, C.: Recumbent bikes—trikes—velomobiles: An analysis of (single vehicle) crashes. In: Proceedings, International Cycling Safety Conference 2015, 15–16 Sept 2015, Hannover, Germany, pp. 25 (2015)

24. Corben, B., van Nes, N., Candappa, N., Logan, D. B., Archer, J.: Intersection study—Task 3 report. Monash University Accident Research Centre, Report No. 316c. http://www.monash.edu/__data/assets/pdf_file/0006/217617/muarc316c.pdf (2010)
25. Garber, N. J., Gadirau, R.: Speed variance and its influence on accidents. In: AAA Foundation for Traffic Safety, Washington DC (1988)
26. Velometro Mobility Incorporated: https://www.velometro.com/veemo/ (2020). Accessed 29 April 2020

The UK Approach to Greater Market Acceptance of Powered Light Vehicles (PLVs)

Huw Davies, Allan Hutchinson, Richard Barrett, Tony Campbell, and Andy Eastlake

Abstract This paper summarises the UK activity for powered light vehicles (PLVs) with the purpose of driving market acceptance. If alternative vehicle concepts are to emerge from the margins and transition into the main stream, there is a need to think differently. This opportunity has motivated a number of UK organisations to come together as a working group and identify a way forward. We contend that thinking differently requires a reshaping of the whole value chain. Each of the partners has contributed to this activity and we describe the development of a pathway towards the realisation of a UK PLV market. Research and policy development requirements for the UK market are defined, supported by a discussion on two specific segments of the PLV market—light freight vehicles and micromobility.

Keywords Powered light vehicles · UK transport policy · Freight vehicles · Micromobility · Transition pathways

H. Davies (✉)
Institute for Future Transport and Cities, Coventry University, Coventry, UK
e-mail: ac2616@coventry.ac.uk

A. Hutchinson
School of Engineering, Computing and Mathematics, Oxford Brookes University, Oxford, UK

R. Barrett
School of Engineering, Liverpool University, Liverpool, UK

T. Campbell
Motor Cycle Industry Association (MCIA), Coventry, UK

A. Eastlake
Low Carbon Vehicle Partnership, London, UK

© The Author(s) 2021
A. Ewert et al. (eds.), *Small Electric Vehicles*,
https://doi.org/10.1007/978-3-030-65843-4_4

1 Introduction

In the UK, surface transport is the largest sector contributor to GHG emissions [1], urban road speeds have reduced by almost 5% in the past 4 years to 18.7 mph [2], whilst nine out of ten Londoners say air pollution is at crisis level [3]. In responding to the problems of climate change, congestion and air quality, it is increasingly recognised that there is a need to re-evaluate our transport options and energy sources for them.

Smaller, lighter, and more energy efficient vehicles are seen as an alternative to traditional mobility options, especially where a move towards other existing forms of transport or a reduction in transport provision is not a viable alternative. The dichotomy is that these alternative concepts are currently perceived as inferior to traditional mobility solutions, resulting in marginalisation and disbenefits to the supplying industry. Further, those vehicle concepts that are more radical in approach also have the issue of public perception to overcome, including functionality issues and concerns about safety.

Transforming the individual vehicle we drive today will require several changes in a holistic sense:

- electric variants;
- the creation of an environment whereby smaller more efficient vehicles can provide a defined benefit;
- changing society to see the value of the most appropriate vehicle for a particular journey;
- supporting suppliers to invest in the development and manufacture of alternative smaller vehicles.

It is the purpose of the UK PLV Working Group to support these changes and provide a forum for co-ordination of activities in this area.

2 Background

Within the UK, and when compared to a number of EU member states, the market penetration of smaller vehicles—those below the M1 passenger car classification—has been limited. Collectively, this category of vehicles is referred to as L-category and consists of powered 2 and 3 wheel vehicles, quadricycles, and microcars. This includes motorbikes and scooters which account for the largest part of the current L-category market.

Although there are relatively few L-category vehicles compared to M and N categories (passenger and goods vehicles) in the UK, they are much smaller, lighter, take up less road space and offer innovative alternatives for mobility, particularly in cities. This engenders the L-category to those concerned with finding solutions to the current externalities around road transport, in particular the pressure on cutting energy consumption, improving local air quality and reducing local congestion.

Recognising this, in 2015, a study into the potential of the L-category market for the UK automotive industry was initiated. This study linked the LowCVP, the UK DfT, and six UK universities: Loughborough, Cardiff, Coventry, Oxford Brookes, Queen's Belfast, and Warwick. The study was broad ranging and recognised that the issue of L-category vehicles was a combination of producer, consumer, and societal interests. Hence, the impacts and benefits were equally wide ranging and considered: air quality and GHG; economic activity including cost to the industry as well as the consumer; and public health including road safety. The result of this study was the launch of the report "Powered Light Vehicles: Opportunities for Low Carbon 'L-Category' Vehicles in the UK" in 2019 [4].

The main conclusion of the Consortium was that PLVs—with either an electric or highly efficient internal combustion engine powertrain—create an opportunity to provide an important contribution to reduction in polluting emissions and energy consumption, both during the manufacture and subsequent operation of these vehicles in the UK. Furthermore, PLVs offer potential growth opportunities for the UK industry, building on the existing engineering capabilities of the automotive sector and, in particular, the motorcycle, motorsports, and niche vehicle sectors, which are well-positioned to exploit opportunities for this category of vehicle. The PLV consortium assessment also identified a number of challenges to the manufacture and use of PLVs in the UK. These relate to the delineation of regulations relating to sit-on and sit-in vehicles, and the incorporation of PLVs into existing UK and EU policy frameworks. The full series of reports written by consortium members are available for further background information and can be downloaded from the LowCVP website: www.LowCVP.org.uk/PLV.

Further to the above activity, the UK Motor Cycle Industry Association (MCIA) has also played an active role in support of the PLV sector in the UK. The MCIA is the Trade Association representing the PLV industry (also known as L-Category vehicles) and has created this strategy to highlight the benefits of incorporating PLVs into the transport mix. As part of its remit, and summarised in the joint publication with the LowCVP "*The Route to Tomorrow's Journeys: Powered Light Vehicles – Practical, Efficient & Safe Transport for All*", are the results of a number of studies investigating the benefits to the UK of supporting the PLV sector [5]. The benefits of these have been explained in full, including the impressive results of the congestion impact and air quality study. Unsurprisingly, the results showed that with a greater shift to electric PLVs (L-category vehicles that have battery electric drivetrains), fewer NO_X emissions will be recorded (from 0.48 kg/day for the chosen baseline scenario to 0.42 kg/day with a 10% modal shift). Reductions were also seen in the scenarios showing the modal shift for PM_{10} and $PM_{2.5}$ (from 0.093 to 0.087 kg/day and 0.051 to 0.048 kg/day, respectively). Emissions relate to the carbon intensity of the electricity grid, and the increasingly renewable content of the UK grid mix (50% in 2019 [6]) means that considerable further reductions may be anticipated in future. Furthermore, the modelled scenarios also found that congestion levels decreased, resulting in reduced delays and shorter journey times for everyone across nine real world junctions (a 20% modal shift gave an 18.4 s reduction in delay averaged across all junctions improving to 39.8 s for a 50% modal shift).

Realising the joint interest in the promotion of PLV in the UK, the MCIA partnered with LowCVP in the UK launch of the *"The Route to Tomorrow's Journeys: Powered Light Vehicles – Practical, Efficient & Safe Transport for All"* and *"Powered Light Vehicles: Opportunities for Low Carbon 'L-Category' Vehicles in the UK"* at the London Transport Museum in 2019. The launch was in response to the UK Government publication of the *"Future of Mobility: Urban Strategy"* [7]. This strategy document recognised that the rise of motor transport has brought substantial benefits, the strategy document also warned that high levels of private car ownership and use have also brought serious challenges. In response the vision articulated in the strategy:

> cleaner transport, automation, new business models and new modes of travel [that] promise to transform how people, goods and services move.

To support that UK Government vision and to position PLV as part of that vision, a new UK PLV working group was established in 2019. The working group is chaired by Coventry University with the MCIA as the secretariat and has the LowCVP, Oxford Brookes University and Liverpool University as permanent members. This group looks to take forward the previous recommendations and discussion points from both the PLV consortium and the MCIA activities. The group has set itself the following mission statement:

> Movement of individuals and goods facilitates production and trade, enhances labour mobility and provides customers with access to goods. Reducing transport-related climate emissions will require transformational changes in thinking, policy, technology and investment to enact. Lighter and smaller vehicles are a key part of the solution (powered light vehicles – PLVs), but struggle to find UK market acceptance. Developing collaborations in research and policy development is a timely intervention to support this.

3   Path to System Transformation

PLVs can play a valuable role in the local transport mix. The recent UK MCIA policy document The Route to Tomorrow's Journeys, refers to a modal shift away from single-occupancy cars and lightly laden vans, towards transport modes that take up less road space and use less energy. The key is to encourage the use of 'the right vehicle for the right journey' and to support transport users in making appropriate choices. However, this transition to the right vehicle for the right journey has proven difficult to realise. It is the inter-relationships between the different parts of the mobility system that can create a resistance to change, resulting in customer lock-in, for example, the economies of scale resulting from the mass adoption of one type of transport modality can cascade through the system reinforcing the existing support structures and locking out modalities that rely on alternative structures. It is understanding and managing these relationships, or how component changes impact upon them, that is key to transformation of our mobility system. For PLVs to move from niche to mainstream, acceptance will require concurrent changes in different parts of the system.

Before alternative mobility solutions, such as the wide spread adoption of PLVs, become widely accepted and economically competitive, they are required to overcome such barriers as technology inter-relatedness, vested interests, legal frameworks that fit the use of historical and present technologies, and the limitations imposed by the evolved consensus over how technology should be designed limiting the vision of business and governments [8]. Existing regulation for PLV application is complex and because traffic laws are defined on national level, different rules are applied across European member states and the UK. Driving licences, for example, are regulated on EU level, but exceptions are permitted for member states for some parameters, for example minimum driving age [9]. Based on this understanding, an exclusively technology-focused approach is unlikely to provide the change required, but it can be an important component.

In response to the research challenge, the PLV Working Group seeks to engage the engineering with the social, economic and planning aspects of the market. The objective is to use a multidisciplinary approach to understand the risk to investment in the emerging PLV value chain in the UK and in the EU. This understanding will inform the development of a framework that will support and strengthen the formulation of technology and public policy to reduce investment risk and strengthen the development, deployment, and diffusion of PLV technologies.

Policy instruments themselves may address a number of market imperfections. However, the way in which national and local governments use these instruments is highly contextualised. First, there is an enormous diversity of policy measures with regards to PLVs in the UK. This alone is indicative of the fact that there is great uncertainty over what works. Second, in terms of socio-technical transitions, the underlying imperative is to stimulate the change process, and yet the uncertainty over policy tools may hinder or even obstruct the change process, because the policy development is primarily evidence based. This presents a number of problems when discussing PLVs. The first is that there is a paucity of evidence upon which to base policy. The second is the variation in vehicle types that exist in different regions due to regulatory frameworks.

It is the inter-relationships between the different parts of the mobility system that can create resistance to change. It is understanding and managing these relationships, or how component changes impact upon them, that is key to transformation of the mobility system. For PLVs to move from niche to mainstream, acceptance will require concurrent changes in different parts of the system. To demonstrate this, two cases are presented from the UK perspective. One, the case of micromobility, is used to demonstrate where existing policy creates customer lock-in, by excluding alternative forms of emerging mobility acceptable to the market place. The other, light freight vehicles, is used to demonstrate where a potential market shift could be supported by appropriate co-ordination of technology and public policy.

3.1 Case 1: Micromobility and the UK

Micromobility is an ambiguous term associated with a rapidly evolving range of powered light vehicles that are increasingly populating streets across the globe (see Fig. 1).

In the United Kingdom and Ireland, these types of motorised micro-vehicles are simply excluded from public roads (and pavements) until the definitions of vehicles permitted for use on the road are updated to include them.

The questions challenging national and local policy makers include:

- Should they be licensed on a technology-specific basis or on a more general mass, power and speed basis?
- Where should they be allowed to operate?
- Will infrastructure need to be adapted to allow for their safe use?
- Which traffic safety requirements should national and local authorities place on shared micromobility operators?

More recently the UK government has opened a call for evidence specifically targeting the micromobility sector. *"The Future of Transport Regulatory Review Call for Evidence"* published in March 2020 asks:

> How, for instance, can e-scooters make life cheaper, more convenient, and maybe a bit more exciting? But also: how safe are they, for their riders and for other road users, and how sustainable? Will they really reduce traffic, or will they reduce walking and cycling more? [9].

Within the UK, the MCIA, a partner in the UK PLV working group, has been active on proposing solutions to the above. Firstly, the specification of the e-scooters used in the UK trials can provide a basis for a framework to categorise e-scooters. Secondly, this framework – and suitably developed according to the findings of those trials – will inform any future legal specification for those wishing to purchase an e-scooter. This approach provides the foundation for making these vehicles accessible—provided that accompanying changes are introduced to

Fig. 1 Examples of the types of vehicle that MCIA believes would come under the scope of micromobility (own visualisation)

licensing [10], training and personal protective equipment (PPE), etc. Further, such an approach also provides an additional level of certainty to the supplying industry that leads to increase in product availability and diversity.

3.2 Case 2: Powered Light Vehicles for Freight and the UK

Receiving goods is an essential component of business, but these deliveries are also a large contributor to congestion and air pollution. This is acknowledged as part of the UK Government's Future of Mobility: Urban Strategy [7]:

> There is significant potential for new modes of transport to replace traditional vehicle miles in urban areas. This could alleviate congestion, reduce noise and emissions, and improve traffic flow. For example, trials of electric cargo bikes showed that they have the potential to increase road speeds in congested areas as well as reducing emissions, costs and delivery times when compared to van-based last mile delivery

> Innovation that supports the more efficient movement of goods, for instance through the use of consolidation hubs or freight brokerage platforms matching goods and vehicle space, will also be important to reduce congestion.

The advantages of using PLV, for example, electric cargo bikes, in arriving at a clean and safe system for the delivery of goods should be fully explored and considered as part of a wider transport strategy and local authority provision. However, the reality is slightly different. Within the UK, a PLV freight sector remains an enigma when compared to the wider light commercial vehicle (LCV) fleet. To understand this underrepresentation requires a closer examination of the finer nuances of how a PLV freight sector would function in the UK. For any new mobility concept, the requirement is

- that it finds alignment with policy;
- that it is economically viable;
- that is accepted by society;
- that industry has the incentive to invest;
- that it is in step with the environment into which it is deployed;
- and that ambiguity over its legal status is addressed.

Concepts that promote a new technology, but fail to consider the economics or the legal status within the market fail to achieve that system transformation. Concepts that promote a new business model, but fail to provide the technology to support that business model, or do not engage with the consumer, fail to achieve that system transformation. Concepts that promote new approaches to manufacture, but fail to consider the wider role of policy in promoting and sustaining existing approaches fail to achieve that transformation. However, the introduction of concurrent concepts that maintain or strengthen inter-dependencies between the various factors can lead to a system transformation. Taking a new vehicle concept and changing the retail environment to account for the different unit cost can succeed. Taking a new mobility concept and changing the regulatory environment can engage with the consumer and succeed.

In the UK context, PLVs used for freight can be considered a new product. Indeed, with their radically different composition (payload, performance, cost base, etc.) and the substantial change, this requires to business systems and accompanying value chains, it can be argued that PLVs would also constitute a new industry. In this context, success requires that technology change has to be partnered by a change to business models, and of course needs customer acceptance. This has higher risk and generates resistance to change.

In response, the acceptance and the success of PLVs for freight will depend strongly on identifying effective use cases that provides user friendly and economic viable operation. New policy and legislative measures can then be introduced that will support the provision of these use cases. For example, the introduction of special operational specifics of L-category traffic regulations like use of bicycle lanes, pedestrian zones or reverse one-way driving. These are presently excluded in the UK for most of the PLV solutions that would be suitable for the type of freight deliveries envisaged by the UK Government strategy document. Further, enabling such special operational specifics, and ensuring there is a level of stability associated with these changes, would also have the benefit of supporting the freight sector in identifying and quantifying risks with new business models that can inform its business planning and forward investment. Factors influencing economic operations include appropriate licensing arrangements, financial incentives and operational motivations such as privileged access, low running costs, and so on.

At present, business models around the use of PLV for freight are generally insufficient, which restrict the development of this sector outside of their core (niche) markets. Current and evolving policy frameworks, that enable special operational specifics, will speed up the adoption and expansion of the PLV for freight. To support this requires regulatory change, which will have the outcome of creating new market spaces. The UK PLV Working Group will be active in this area.

4   Discussion

Replace existing modalities with PLVs has the potential to generate a range of positive outcomes, from clean air to reduced congestion. Hence, there is strategic value in the support of PLVs and this has been recognised in the UK strategy for urban mobility. However, with such a complex system as road transport, the interventions in support of a move to PLVs must be planned and must be co-ordinated in order to support wider market acceptance. This paper has summarised the UK activity for PLVs with the purpose of driving market acceptance.

It has also been shown how the activity links to the UK strategy for future urban mobility. Further to the above, a result of this activity has been the identification of a number of key recommendations and discussion points. These are summarised below:

Key recommendations:

- Raise awareness among key stakeholder within the UK value chain
- Undertake whole life-cycle assessments of PLVs
- Make representations at EU level to include PLVs in fleet averages,
- Implement technical R&D projects needed to optimise systems applicable PLVs
- Conduct UK focused end-user research to build on the insights from early adopters
- Re-assess UK low-carbon vehicle purchase incentives to include PLVs.

Discussion points:

- Currently only M, N, and O category vehicles are supported by the National Small Series Type Approval (NSSTA) in the UK. To assist small UK companies in entering the PLV market, the UK Department for Transport (DfT), along with the Vehicle Certification Agency (VCA), is to consult stakeholders on creating a NSSTA for PLVs.
- Improve regulatory delineation between open "sit-on" and enclosed "sit-in" L-category powered light vehicles. The former (scooters, motorcycles, quad-bikes) are already an established market, whereas the latter has neither a clear identity for consumers (e.g. safety standards) nor the best legal framework (e.g. suitable drive cycle/crash test procedures) to flourish.
- New regulations are needed for a minimum safety cell performance: frontal and side impact crash tests performance (appropriate for the class), rollover test, seat with whiplash protection. Impact test for pedestrians (enhancement of existing functional safety regulations). Further accident research performed on L-category PLVs to review and amend the initial test proposed for the frontal, lateral, and roll-over load cases.
- Funded research is undertaken to ensure that future vehicles of all sizes are able to detect PLVs in their AEB sensing algorithms. Along with funded research in the development of AEB, city, inter-city and pedestrian safety and funded research into the implementation of an integrated safety test protocol of L-category PLVs.

To deliver against these recommendations and to support ongoing discussion, a UK PLV working group has been established. To date the working group has focused on two emerging PLV sectors: micro-mobility and freight.

- For micro-mobility, an emerging consumer demand is at present tempered by a gap in the classification of the product. A recommendation for classification that extends the existing L-category topology is proposed. Adoption by Government would enable the appropriate discussion around category definition—taking into account concerns around safety in addition to congestion and emission benefits—which would in turn lead to increased consumer confidence and industry participation.

- For freight, the issue is that the business models around PLVs are generally insufficient outside of specific niche applications. The adoption of PLVs will depend strongly on expanding the effective use case through a combination of product innovation and creating new market spaces through a change in surrounding conditions, such as creating specific regulations like use of bicycle lanes, pedestrian zones or reverse one-way driving. An effective use of these specific regulations may be viewed as a key enabler for new delivery services.

5 Next Steps

Following on from the recently announced UK government call for evidence on transport regulation and specifically targeting the micromobility sector [9], a further publication of 26 March 2020 "*Decarbonising Transport, Setting the Challenge*" [11] states:

> We will use our cars less and be able to rely on a convenient, cost-effective and coherent public transport network. From motorcycles to HGVs, all road vehicles will be zero emission. Technological advances, including new modes of transport and mobility innovation, will change the way vehicles are used. Our goods will be delivered through an integrated, efficient and sustainable delivery system

The strategic position of the UK government—outlined in the call for evidence on transport regulation and the decarbonisation challenge—creates a clear opportunity for the development of a robust and optimised PLV market within the future of UK transport and will form a continuing basis for further coordination of the UK PLV group.

References

1. Committee on Climate change: Reducing UK emission—2019 progress report to parliament. https://www.theccc.org.uk/publication/reducing-uk-emissions-2019-progress-report-to-parliament/ (2019). Last Accessed 6 Nov 2019
2. Department for Transport: Transport Statistics Great Britain 2017, https://assets.publishing.service.gov.uk/government/uploads/system/uploads/attachment_data/file/664323/tsgb-2017-print-ready-version.pdf (2017). Last Accessed 20 Sep 2018
3. London.Gov: 9/10 Londoners say air pollution is at 'crisis' levels. https://www.london.gov.uk/press-releases/mayoral/londoners-poll-air-pollution-is-at-crisis-levels (2017). Last Accessed 13 Nov 2017

4. LowCVP: Micro vehicles: opportunities for low carbon 'L-Category' vehicles in the UK. https://www.lowcvp.org.uk/assets/reports/LowCVP_Powered_Light_Vehicles_2019. pdf&usg=AOvVaw0MLVstf3OX_PqdfrhtvPwC (2019). Last Accessed 1 May 2020
5. MCIA: The route to tomorrow's journeys: powered light vehicles—practical, efficient & safe transport for all. https://mcia.co.uk/attachment/ead15463-d738-4dd0-a43d-bc48ca633fb7 (2019). Last Accessed 1 May 2020
6. Raugei, M., Kamran, M., Hutchinson, A.: A prospective net energy and environmental life-cycle assessment of the UK electricity grid. Energies **13**(9), 2207 (2020)
7. DfT: Future of mobility: urban strategy. https://assets.publishing.service.gov.uk/government/ uploads/system/uploads/attachment_data/file/846593/future-of-mobility-strategy.pdf (2019). Last Accessed 1 May 2020
8. Sanden, B.A., Azar, C.: Near-term technology policies for long-term climate targets— economy wide versus technology specific approaches. Energy Policy **33**, 1557–1576 (2005)
9. DfT: Future of transport regulatory review: call for evidence. https://assets.publishing.service. gov.uk/government/uploads/system/uploads/attachment_data/file/873363/future-of-transport- regulatory-review-call-for-evidence.pdf (2020). Last Accessed 5 May 2020
10. Directive 2006/126/EC of the European Parliament and of the Council of 20 December 2006 on driving licences
11. DfT: Decarbonising transport: setting the challenge'. https://assets.publishing.service.gov.uk/ government/uploads/system/uploads/attachment_data/file/878642/decarbonising-transport- setting-the-challenge.pdf (2020). Last Accessed 5 May 2020

Case Studies and Applications of SEVs

The ELVITEN Project as Promoter of LEVs in Urban Mobility: Focus on the Italian Case of Genoa

Francesco Edoardo Misso, Irina Di Ruocco, and Cino Repetto

Abstract One of the growing innovations in the electric vehicle market concerns light electric vehicles (LEVs), promoted at local and national level by many initiatives, such as the European project ELVITEN, involving six cities, which is analysed in the present paper in relation to the Genoa pilot case study. In Italy, LEVs have been increasingly successful, as the number of their registrations shows (+76% in 2019 compared to 2018). In this context, the city of Genoa, where a considerable fleet of mopeds and motorcycles (214,499 in its metropolitan area in 2018) circulates, lends itself well to the experimentation of two-wheeled LEVs. The monitoring of the use of LEVs within the framework of the ELVITEN project has shown that the average daily round trips recorded in the metropolitan area of Genoa are equal to 15–20 km, thus reinforcing the idea that LEVs represent a valid alternative to Internal Combustion Engine (ICE) private vehicles. Moreover, the characteristics of the travel monitored and the users' feedback highlight that the question of range anxiety is less present than expected. Finally, and contrary to our expectations, the data analysis indicates that the use of LEVs in Genoa during two months of Covid-19 pandemic lockdown—March and April 2020—shows a decrease of 21%, while the average decrease recorded by the six cities globally considered is 51%.

Keywords LEVs · E-mobility · Genoa

1 The ELVITEN Project

The new frontiers of mobility include a number of new transport models, starting from the use of electric vehicles, micro-mobility or soft mobility systems up to electric self-driven vehicles, allowing environmentally friendly transit. Such models, which play an important role in the smart mobility scenario, require a solid

F. E. Misso (✉) · I. Di Ruocco · C. Repetto
T Bridge S.p.A, Via G. Garibaldi 7/10, 16124 Genoa, Italy
e-mail: f.misso@tbridge.it

© The Author(s) 2021
A. Ewert et al. (eds.), *Small Electric Vehicles*,
https://doi.org/10.1007/978-3-030-65843-4_5

digitalisation. In this line, many cities have adopted new transport systems based on electric mobility, encouraging the use of light electric vehicles (LEV). The ELVITEN project[1] (Electrified L-category Vehicles Integrated into Transport and Electricity Networks) aims at the diffusion and simultaneous integration of LEVs (e-bicycles, e-scooters, e-tricycles and quadricycles) in order to offer a new concept of sustainable urban transport system for people mobility (work, tourism, and leisure) and goods delivery. The ELVITEN project moved its first steps in 2017 thanks to Horizon 2020 funding. The project involves 21 partners, coordinated by the lead partner ICCS (Institute of Communication and Computer Systems) in Athens, including T Bridge that has the role of the coordinator for developing the ICT tools supporting the experimentation.

The ELVITEN project methodology proposes, implements, and demonstrates innovative deployment schemes to address in a comprehensive way all major issues related to the market entry of LEVs. These issues are poor user awareness due to a low level of information on vehicle performance and functionality, consumer concerns such as range anxiety, and inadequate mobility planning for LEVs of existing paid infrastructure networks and inadequate planning for the integration of LEVs into urban transport networks.

The approach of the project includes the preparation, usage, and analysis phases, analysing the mobility demand in six pilot cities (Berlin in Germany, Bari, Genoa and Rome in Italy, Malaga in Spain, and Trikala in Greece) and the identification of the usage schemes and potential number of users on the basis of each city's mobility conditions. In order to create profiles of potential ELVITEN users in each city, the project provides a focus on the following clusters of LEV users: systematic urban commuters, occasional urban travellers, and light delivery companies.

The actions are carried out in the demonstration cities and are related to the use of vehicles (e.g. parking lots or charging areas) focusing on the most strategic areas such as covered areas close to, for example, railway stations or shopping centres.

The ongoing activities include the set-up of a platform and applications for mobile devices through which citizens can, among others, access ELVITEN services, book charging points, and use sharing services in an integrated logic.

To summarise, the ELVITEN project aims at:

- testing how to enhance/reinforce electric vehicle use in the six pilot cities, reducing pollutant emissions as well as traffic congestion, targeting both cars and light vehicles (motorcycles, pedal-assisted bicycles, tri- and quadricycles);
- increasing LEV usage in urban areas, through the development of replicable usage schemes;
- deploying LEV innovative parking and charging services, to facilitate LEV owners;
- empowering LEV sharing and rental services;
- developing support ICT tools to motivate the usage of LEVs instead of ICE vehicles;

---

[1]www.elviten-project.eu/it/ [1].

- understanding which factors can favour (or discourage) users' choices towards such kinds of vehicles.

The main purpose of the present paper is to show, in the light of the ELVITEN experience carried out in the pilot city of Genoa—which has 575,577 inhabitants [2] and is squashed between the Apennines and the Ligurian Sea—that LEVs are a viable and sustainable alternative, both for individual use and for sharing. We also focus on the observation of market trends in LEV trade and on initiatives related to the promotion of electric mobility such as the design of parking lots and parking areas for LEVs.

2　The Current Trend in the LEV Sector

Since 2017, there has been a growing diffusion of LEVs in Europe. In Italy, for example, this trend was demonstrated by the increase of the registrations of all categories of electric vehicles (EV), with a growth of 104% in 2018 compared to the previous year (see Fig. 1) [3, 4].

Italy, as other countries in Mediterranean Europe [5], is characterised by a massive use of two-wheeled vehicles (motorcycles, mopeds). In the first half of the year 2019, an increase of 135% and of 64% regarding the registrations of electric motorcycles (1,163 units) and e-mopeds (2,022 units), respectively, was observed in comparison with 2018 (see Fig. 2) [5].

Another performance that is worth highlighting is the growth in the number of charging points for EVs in Italy: 13,721 in February 2020 (there were 10,647 in September 2019) and 7,203 charging stations (425 in the Liguria Region) accessible to the public (5,246 in 2019) [6]. Of the total number of charging stations, 73% are located on public ground, while 27% are located on private areas open to the public, such as supermarkets or shopping centres [6].

At the national level, the current guidelines to support LEVs in Italy are the *Decreto Ministeriale* (Ministerial Decree) no. 229/2019 [8] that launched the experimentation of electrical micro-mobility, and the *Piano Nazionale Infrastrutturale per la Ricarica dei veicoli alimentati ad Energia elettrica* (PNIRE)

Fig. 1 Thousands of registrations of BEVs (battery electric vehicles) and PHEVs (plug-in hybrid electric vehicles) in Italy [3]

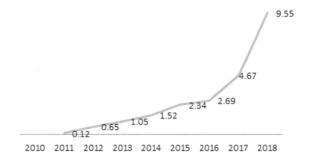

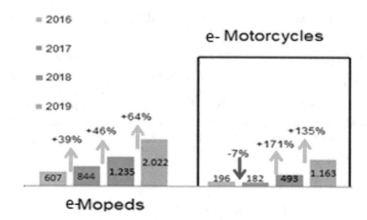

Fig. 2 Trend of the electric two-wheeled vehicles Italian market, 2016–2019 [5]

(National Infrastructure Plan for the Recharging of Electric-powered vehicles) [9] promoted by the Italian Ministry of Infrastructure and Transport (MIT) [7]. At the international level, the landmarks are the European Regulation no. 168/2013, that aims to establish harmonised rules for the type-approval of L-category vehicles, and the Transport 2050 Strategy [10, 11], that outlines a roadmap for a competitive transport sector with ambitious sustainability targets, such as reduction in CO_2 emissions and fleet renewals to reach 100% of electric and hybrid vehicles.

Finally, the increasing use of ICT in sustainable mobility has played an important role in the transport sector promoting the shift from the central role of the hardware component (e.g. vehicles, roads, or infrastructures) towards a mobility model that serves the needs of the users. Therefore, ICT is becoming a strong support for public administrations and for users in the field of electric mobility to integrate EVs and LEVs globally with the transport network, especially in the framework of Mobility as a Service (MaaS). In this light, the MaaS model becomes eMaaS [12] in order to optimise the use of electric vehicles and overcome the disadvantages of "traditional" mobility and "traditional" sharing.

3   The Case Study of Genoa

The city of Genoa is the first large urban area with the highest number of ICE light vehicles in circulation in Italy and has one of the highest ratios of light vehicles per inhabitant in Europe, of over 26 LEVs per 100 inhabitants [13].

Indeed, motorcycles and mopeds make more than 28,000 trips per working day while cars about 47,000. Data on vehicle fleets from the Automobile Club of Italy (ACI) highlight about 397,111 light vehicles (motorcycles and mopeds) in the Liguria Region, 218,066 in the entire metropolitan area of Genoa and 144,970 in the city of Genoa [14].

Still, the use of electric scooters and quadricycles is low, and this is due to the persistence of psychological and technological barriers, such as asymmetric information, range anxiety, and charging operations (ELVITEN project, 2017, also based on previous experiences).

Moreover, the topographic characteristics of Genoa, with considerable differences in altitude and traffic-restricted streets in the historical centre, make this city, among the other European and Italian cities, suitable to be chosen as a pilot case to experiment the spread of electric vehicles in the framework of the ELVITEN project.

Since 2013, the City of Genoa has put in place several actions, which testify to the strong commitment of the city in the field of LEVs. Thanks to projects in which Genoa is the protagonist, these measures include the spread of the network of charging points and the launch of LEVs such as two-, three-, and four-wheel vehicles on the metropolitan territory.

ELVITEN is surely an example of such projects. It follows the legacy of the Ele. C.Tra. "Electric City Transport" project, funded by the IEE programme (Intelligent Energy—Europe) in 2013 and completed at the end of 2015. It included eleven international partners and the three important pilot cities Genoa, Florence, and Barcelona that, with another seven cities around Europe, laid the foundations for the development of light electric vehicles. From the very beginning, the relevance of the stakeholders' networking and the role of physical (e.g. charging points) and immaterial (e.g. policies and raising citizens' awareness) actions to be done was clear.

Finally, the city of Genoa renewed its membership of the Covenant of Mayors in 2018, a pact between eleven cities in order to reduce greenhouse gas emissions by at least 40% by 2030.

3.1 The ELVITEN Project in Genoa

The experimentation with electric vehicles deals with different activities such as the monitoring of the range anxiety, in particular, and users' confidence in LEVs, in general, the question of recharging, the diffusion of sharing and the development of e-roaming and ICT tools.

In Genoa, the project promotes the use of LEVs, thanks to Kyburz, the supplier of ten electric L5e-A vehicles (tricycles), and private suppliers of one e-scooter and six e-quadricycles. Kyburz' tricycles have been conferred to public and private subjects in the Genoa area and are currently being used.

The testing in Genoa involves many ELVITEN partners such as:

- The Municipality of Genoa, which provides free parking for LEVs with the possibility of advanced booking and "bonuses" for virtuous behaviour;
- T Bridge, coordinating technical activities and developing ICT solutions;
- Quaeryon and Softeco, concerning ICT activities and

- Duferco Energy, which provides the charging infrastructure and a dedicated app ("D-Mobility").

The pilot test started in May 2019 and the monitoring of the trips of the demonstration vehicles is still ongoing. For this purpose, each vehicle is equipped with a data logger (blackbox) in order to track every trip start, duration, and route.

The Genoese experimentation includes the implementation in strategic areas of the city of four charging points for two-, three-, and four-wheeled vehicles, which are integrated with the existing infrastructure. This measure promotes the development of a charging network in the city and the incentive for the adaptation of charging sockets in both public and private spaces.

The creation of the hubs has triggered a process of urban transformation, converting existing urban spaces to new functions. Parking spaces have been converted into a multimodal hub promoting inter-modality allowing citizens to use local and/ or private public transport and EVs.

Genoa has also initiated a process of civic involvement by linking mobility to the citizen. A plurality of ICT solutions is currently in phase of testing or development, such as:

- the booking of the charging points in the four Genoa hubs, thanks to the Duferco Energy "D-Mobility" mobile app, which allows the citizen to search, select, and book a charging point available, pay and monitor the charging;
- the monitoring of the EV fleet of tricycles with the blackboxes of the vehicle producer (Kyburz) and the following analysis made by the ELVITEN responsible partners;
- the booking and management of "positive incentives" to be given to citizens for their virtuous behaviours with regard to electric mobility, also to be used in other services proposed by the municipality and integrated in future in a wider rewarding logic for Smart-City citizens. For example, the possibility of offering free tickets for the local municipal museums is currently in progress. The system is implemented by the platform "Myeppi"®,[2] an outcome of the previous European project MOVEUS owned by Genoa Municipality, Quaeryon and Softeco, which has been adapted and customised for the ELVITEN needs.

For each ELVITEN Demonstration City, four types of specific solutions have been defined to carry out the demonstration through SERVICES, ICT TOOLS (e.g. apps, booking and brokering services), POLICIES promoted by public bodies and SPECIFIC COMMUNICATION/PROMOTION INITIATIVES.

The specific expected results, partially implemented, for Genoa are synthesised in Table 1.

---

[2] www.myeppi.com.

Table 1 Solutions deployed in Genoa

Services	Creation of a widespread charging point network via the installation of four e-charging hubs, including covered shelters for electric bicycles and by integrating charging points in private areas, open for public use
	Integration of the charge network with the existing infrastructure
	Offering of ten three-wheelers by Kyburz for free use by local logistics companies
ICT Tools	Brokering service for the shared three-wheelers and for booking of charging points ("D-Mobility" app)
	Management system for the e-charging hubs
	LEV fleet monitoring tool and digital coach app
Policies	Genoa Municipality provides free parking in downtown parking areas for LEVs, with the possibility of advance booking
	"Bonuses" for virtuous behaviour, as collected by the Incentive Management App (platform "Myeppi"®)
Specific communication/ promotion initiatives	Genoa organises a permanent exhibition, in which LEV providers present their products and inform visitors about the state of art of LEVs
	LEVs are offered to citizens for free testing

Source ELVITEN project, elaboration by authors

3.2 Focus on Genoa Data

The ELVITEN trips data (see Fig. 3) were collected before the COVID-19 period, showing the number of trips per month for all pilots up to February 2020. The total number of trips registered is 26,798 trips, of which 532 are active users. These are identified users who also participated in compilation of the questionnaires that were submitted by LEV-users during the monitoring activity of the project. The diagram shows a slow increase in the number of users after January.

The trip data collected by month presents an increasing peak of displacements in October 2019, highlighting that the testing of LEVs progressively increased as the demonstration was carried on. The winter period shows a not surprising decrease in the use of LEVs.

In the framework of the pilot experimentations, almost 1,000 questionnaires were filled out by LEV owners and users of shared vehicles. They highlighted that 88% of users of all pilot cities were satisfied with travelling on their LEVs and pointed out strong confidence in the use of LEVs as an alternative to ICE vehicles.

In Genoa, during the pilot period (up to February 2020, before the COVID-19 pandemic), more than 2,500 trips were made with the monitored ELVITEN LEVs (see Fig. 4) and 260 in March 2020 (see Fig. 5). These intermediate results, which are aligned with the other cities involved despite the poor attitude to use EVs in Genoa, show a good response of the citizens.

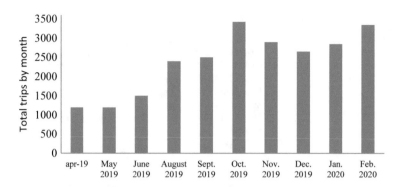

Fig. 3 Total trips per month for all pilot cities. *Source* ELVITEN project, 2020

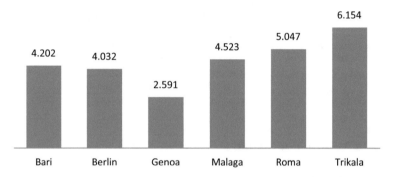

Fig. 4 Evolution by city of the total number of trips monitored (from May 2019 to February 2020). *Source* ELVITEN project, elaboration by authors

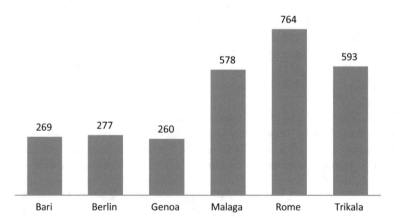

Fig. 5 Evolution by city of the number of trips monitored (March 2020). *Source* ELVITEN project, elaboration by authors

The use of the ELVITEN LEVs in Genoa is characterised by one of the highest numbers of the average distance per trip that are registered in the pilot cities with more than 5.5 km (see Fig. 6).

Due to the geographical nature of the city, many trips are made within the urban area, characterised by a distance of 4–9 km and duration of less than 20 min. However, some vehicles make trips in suburban areas covering a greater distance reaching 25 km. Such a distance is also the major trip catchment area of the urban centre of Genoa (see Fig. 7). In consideration of the LEVs range equal to about 60 km, the assumption that LEVs can replace ICE vehicles in urban and extra-urban areas under specific conditions is verified.

Regarding the average number of trips per user (a sort of utilisation rate) Genoa is peculiar. Indeed, while in most pilot cities sharing services were set up, in Genoa the 17-vehicle demonstration fleet (ten tricycles, one e-scooter and six e-quadricycles) has been assigned to specific users; consequently, we can notice a high average value of trips per user in comparison with other cities (see Fig. 8).

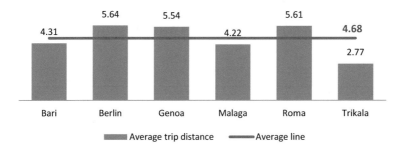

Fig. 6 Average distance per trip by city (up to February 2020). *Source* ELVITEN project, elaboration by authors

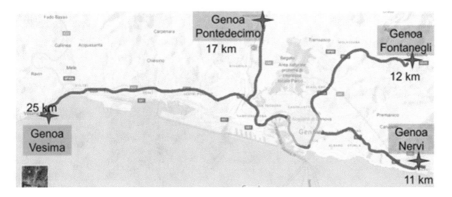

Fig. 7 Trip catchment area in Genoa. *Source* elaboration by authors

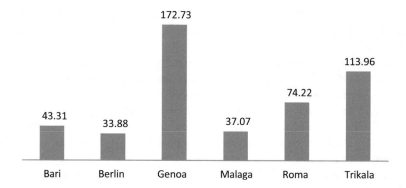

Fig. 8 Average number of trips per users. *Source* ELVITEN project, elaboration by authors

For all pilots, including Genoa, the day timeslot with the greatest number of trips registered is the morning time (10–13 o'clock with a quota of 32%) and secondly the morning rush hours 7–10 o'clock (22%) (see Fig. 9), highlighting the interesting share of the ELVITEN LEV use for leisure and more occasional trips than the usual journeys to get to offices or schools. In addition, the trips are made more often on weekdays (83%) and less on the weekend (17%) (see Fig. 10).

Fig. 9 Trips by timeslot. ELVITEN project, elaboration by authors

Fig. 10 Trips by day of the week. ELVITEN project, elaboration by authors

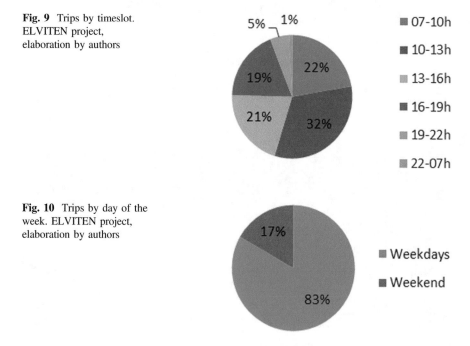

Table 2 Evolution by city of the number of trips per month monitored

City	Average trips/Month (start pilot—Feb 2020)	Trips in March 2020	Trips in April 2020*	Δ % (March–April 2020) (%)
Bari	326	269	5	−98
Berlin	328	277	283	2
Genoa	218	260	206	−21
Malaga	377	578	157	−73
Rome	571	764	295	−61
Trikala	511	593	397	−33
Total	2,331	2,741	1,343	−51

*until 28 April

Source ELVITEN project, elaboration by authors

Finally, the monitoring of the number of trips carried out by LEVs in March and April 2020 (see Table 2) shows that in most of the pilot cities, the travel number decreased but not markedly, despite the spread of COVID-19 and the consequent blockage of travel. The figures for Genoa, in particular, show that despite a 21% decrease, it is lower than the other cities involved (average of 51%).

4 Conclusions: Challenges of LEVs

The first results of the still ongoing ELVITEN project allow us to collect, analyse, and give relevant insights into the use of LEVs in everyday mobility. First, the specific pilot case of Genoa is showing that the consumers' concerns about the feeling of uncertainty with regard to the possible need to re-charge the vehicle during the trips (range anxiety) is less present than expected. Indeed, users who evaluated LEVs positively (88% of total pilot city users who filled in the questionnaires submitted) expressed that today the challenges facing LEVs are mostly related to the three following issues, listed approximately in order of priority from the citizens' point of view: cost, charging infrastructure and range. Only 3% of shared LEV users and 3% of the owners believe that the range anxiety is the most critical challenge to overcome today in order to make possible and convenient the use of LEVs.

The evidence that the range anxiety is not the first challenge to be tackled in an urban context, such as Genoa, also concern the distance covered per trip, which, as previously highlighted, in Genoa is relatively higher than those usually covered by such vehicles (an average of 5.5 km, with trips reaching 25 km).

The results of the still ongoing ELVITEN experience are showing that working towards putting in place an integrated system of charging infrastructure (in Genoa by four new charging hubs in the most attractive points of the city) and incentives (set of policies offered and reward systems for the users) supports the use of LEVs.

In this light, the characteristics of LEV trips, in terms of, for example, number, distance covered, or utilisation rate, are not strongly affected by the impacts of the constraints due to the COVID-19 pandemic.

Acknowledgements This paper and the research behind it would not have been possible without the participation of T Bridge in the ELVITEN project. We thank our project partners (https://www.elviten-project.eu/en/about/) for allowing us to use the project data to produce this paper. We would also like to thank our colleagues Michele Solari and Simone Porru from T Bridge, the Duferco Energy project partner, who have undoubtedly contributed to the interpretation of the data in this paper.

Funding for this research was provided by the HORIZON 2020 project "ELVITEN".

This project has received funding from the European Union's Horizon 2020 research and innovation programme under grant agreement No 769926.

References

1. ELVITEN homepage: https://www.elviten-project.eu/it/. Last Accessed 22 April 2020
2. ISTAT homepage: http://dati.istat.it. Last Accessed 22 April 2020
3. Motus-e Homepage: (2019). https://www.motus-e.org. Last Accessed 22 April 2020
4. The European House Ambrosetti, Motus-e (in collaboration with): La filiera della mobilità elettrica "Made in Italy". (2019). https://www.motus-e.org/wp-content/uploads/2019/07/1907_Executive-Summary_La-filiera-e-Moblity-in-Italia.pdf
5. ANCMA Homepage: http://www.ancma.it/statistiche/. Associazione Nazionale Ciclo Motociclo e Accessori. Last Accessed 22 April 2020
6. Motus-e.: Le Infrastrutture di ricarica pubbliche in Italia (2020) https://www.motus-e.org/wp-content/uploads/2020/03/Report-IdR_Marzo_MOTUS-E.pdf
7. MIT Homepage: http://dati.mit.gov.it. Last Accessed 22 April 2020
8. Il Ministro delle Infrastutture e dei Trasporti: (2019) M_INF.GABINETTO.REG_DECRETI. R.0000229.04-06-2019. http://www.mit.gov.it/sites/default/files/media/notizia/2019-06/schema%20DM%20micromobilit%C3%A0%20229_2019%20%283%29.pdf. Last Accessed 24 Aug 2020
9. Il Ministro delle Infrastutture e dei Trasporti: (2012) Piano Nazionale Infrastutturale per la Ricarica die veicoli alimentati ad energia Elettrica. http://www.governo.it/sites/governo.it/files/PNire.pdf. Last Accessed 24 Aug 2020
10. Regulation (EU) No 168/2013 of the European Parliament and of the Council of 15 January 2013 on the approval and market surveillance of two- or three-wheel vehicles and quadricycles. https://eur-lex.europa.eu/legal-content/EN/ALL/?uri=CELEX%3A32013R0168. Last Accessed 24 Aug 2020
11. European Commission: Transport 2050: Commission outlines ambitious plan to increase mobility and reduce emissions. (2011). https://ec.europa.eu/commission/presscorner/detail/en/IP_11_372. Last Accessed 24 Aug 2020
12. Reyes Garcia, J.R., Lenz, G., Haveman, S.P., Bonnema, G.M.: State of the art of electric Mobility as a Service (eMaaS): an overview of ecosystems and system architectures. In: 32nd International Electric Vehicle Symposium (EVS32) (2019)

13. ACI Homepage: (2019) http://www.aci.it/ ACI Automobile Club d'Italia. Last Accessed 22 April 2020
14. Automobile Club d'Italia: Autoritratto 2019. (2019) http://www.aci.it/laci/studi-e-ricerche/dati-e-statistiche/autoritratto/autoritratto-2019.html. Last Accessed 03 July 2020

Small Electric Vehicles in Commercial Transportation: Empirical Study on Acceptance, Adoption Criteria and Economic and Ecological Impact on a Company Level

Tim Hettesheimer, Cornelius Moll, Kerstin Jeßberger, and Saskia Franz

Abstract Small electric vehicles (SEVs) in commercial transportation have the potential to reduce traffic and its impacts, especially in urban areas. Companies, however, are still reluctant to implement SEVs. Therefore, the aim of this contribution is to shed light on the acceptance of motives for and obstacles to the use of SEVs in commercial transportation. Since the use of SEVs is often discussed in the context of innovative city logistics concepts, such as micro-hubs, our aim is also, to explore the acceptance, economic, and ecological potentials of SEVs in combination with micro-hubs. We use a multi-method approach and combine an online survey with in-depth interviews as well as a total cost of ownership (TCO) and CO_2 calculation. Analyzing 350 responses to an online survey revealed that around half the companies surveyed have no knowledge of SEVs. This implies high unexploited potential, since 25% of these companies can imagine using them. In-depth interviews with logistics service providers (LSPs) or logistics departments from different sectors revealed that six of the 13 interviewed LSPs would be willing to implement this concept.

Keywords SEVs · Commercial transportation · TCO · CEP

T. Hettesheimer (✉) · C. Moll
Fraunhofer Institute for Systems and Innovation Research ISI,
Breslauer Straße 48, 76139 Karlsruhe, Germany
e-mail: tim.hettesheimer@isi.fraunhofer.de

K. Jeßberger
Julius-Maximilians-Universität Würzburg, Sanderring 2, 97070 Würzburg, Germany

S. Franz
Karlsruhe Institute of Technology (KIT), Kaiserstraße 12, 76131 Karlsruhe, Germany

© The Author(s) 2021
A. Ewert et al. (eds.), *Small Electric Vehicles*,
https://doi.org/10.1007/978-3-030-65843-4_6

1   Introduction

Urbanization, economic growth, and structural economic changes have led to a steady increase in road transportation, especially in urban areas [1]. Approximately, 50% of urban traffic is caused by commercial vehicles [2]. This results in heavy congestion, air pollution, and greenhouse gas emissions in urban areas [3, 4]. When considering goods transport, it is striking that this only accounts for between 20 and 30% of total urban traffic, but is responsible for 80% of inner-city traffic jams during rush hours, for example, due to the double parking of delivery vehicles [5]. This situation could worsen in the future. Between 2000 and 2017, the number of courier, express, and parcel deliveries (CEP) in Germany doubled from 1.7 to 3.4 billion and, according to estimates, these are expected to increase further to 5 billion deliveries by 2025 [6]. In this context, the potential use of SEVs in commercial transportation and in the CEP industry in particular is a subject of increasing interest [5, 7]. SEVs are much smaller compared to conventional delivery vans or trucks and require less space while in traffic or parking. Furthermore, they are lighter, more agile and, when being powered electrically, more energy efficient than conventional vehicles. Consequently, SEVs have the potential to reduce traffic and its impact, especially in urban areas.

While there are already a number of SEVs on city streets in private transport, they are still the exception for commercial transport services. The total user potential for commercial transport in Germany has already been estimated, for example, by Brost et al. [7], using data from the mobility survey "Mobility in Germany (MiD) 2017". However, this analysis based on so-called "regular professional trips" only allows an estimation of the maximum technically possible potential (such as driving range or load volume). Furthermore, it is important to note that company-specific criteria have a decisive influence on the user potential [7]. This contribution therefore focuses on a company-specific analysis of user potential. For this purpose, the results of a survey of more than 350 companies and 14 company-specific interviews are combined. This combination of the survey and the interviews allows for a cross-company view as well as a company-specific perspective on acceptance, potentials and obstacles for the use of SEVs in various commercial applications. This approach supports the aim of this contribution to assess the acceptance and company-specific reasons for or against the use of SEVs in commercial transport in general. In addition, due to the high relevance of CEP service providers for efficient urban traffic, an in-depth analysis was conducted on the acceptance and willingness of logistic service providers (LSPs) or logistics departments to use innovative concepts such as SEVs in combination with micro-hubs and on how this concept could contribute to the profitability and environmental impact of innovative city logistics concepts.

SEV definition. In this study, we showed the companies different SEV classifications in advance to give them an impression of the different characteristics of the individual classes. This was done using the following exemplary models: KEP10

(L1e) as a two-wheeled cargo bike, three- and four-wheeled cargo bikes such as the Cargo Cruiser (L2e) or Loadster, the Paxter (L6e), and light electric vehicles such as Twizy Cargo and Microlino (L7e).

2 Materials and Methods

For the analyses of the general potential, an online survey was conducted of 4,000 companies on the distribution and applicability of SEVs in companies. The geographical focus is the "Technologieregion Karlsruhe".[1] Companies of varying size from different sectors of industry were surveyed. The analysis covers the responses of more than 350 companies (return rate of $\sim 7.5\%$).

In addition, interviews were conducted with 14 companies from industries that are considered the most suitable for SEVs in the literature. This involves installation service (1 company interviewed), painting trade (3), chimney sweeps (2), cleaning service (1), nursing service (1), CEP service (1), pharmacy (2), internal factory traffic (1), and delivery service (2). In some sectors, several companies were interviewed to find out whether and how the situation differs between companies in the same sector.

To answer the question about the acceptance of SEVs in combination with micro-hubs and how this concept could contribute to the profitability of innovative city logistics concepts, 13 semi-structured interviews were conducted with experts from different LSPs or logistics departments, supplemented by a total cost of ownership (TCO) calculation, an environmental analysis of CO_2-emissions and a processing time analysis. In the interviews, besides evaluating the acceptance of LSPs to use innovative city logistics concepts, we derived logistics and economic data regarding the current transportations structures of the LSPs. The interview consisted of almost 50 questions. The data derived were used to calculate the TCO, processing times, and environmental impact (CO_2 emissions) of combining SEVs with micro-hubs.

When applying the TCO method, all relevant processes and procedures of a current and a future scenario or concept has to be analyzed and respective costs have to be specified. Since our interviews showed, that the CEP service industry is most likely to implement micro-hubs with cargo bicycles, we assumed that the original distribution structure is transformed to a micro-hub concept. In the status quo, a conventional delivery van starts the distribution of parcels to customers from a regional distribution center outside of an urban center. In the new concept, a battery-electric heavy-duty truck (HDT) transports two swap bodies (see Fig. 1), used as micro-hubs, from the regional distribution center to an urban center, while

---

[1]The region around the city of Karlsruhe, which is an industrial and technological center in the southern part of Germany, comprising approximately 3,240 km² and 1.3 million inhabitants.

Fig. 1 Exemplary representation of swap bodies used as micro-hubs. *Source* Dachser Neuss Swap Bodies. Licensed under CC BY-SA3.0 (https://creativecommons.org/licenses/by-sa/3.0/deed.de). Cut out. Original version: https://de.wikipedia.org/wiki/Datei:Dachser_Neuss_Swap_Bodies.JPG

battery-electric three-wheeled cargo bicycles distribute the parcels from the swap bodies (micro-hubs) to the end-customers.

We considered the following costs, associated with the two logistics concepts: vehicle costs (fixed costs: annual depreciation, insurance costs and taxes; variable costs: fuel or electricity costs and costs for repair, maintenance, and tires), driver personnel costs, equipment and location costs for the micro-hubs and charging infrastructure costs for electrically powered vehicles. The calculations are based on the data collected in the semi-structured interviews. Missing data was collected through further literature and desk research. The logistics parameters used can be found in Table 1 and the techno-economic parameters in. Tables 2 and 3. Based on these data, we set up an average use case for each of the two logistics concepts and we retrieved average vehicle types, route lengths, break and working times, number of parcels to be delivered, etc. All costs, annually-fixed or variable, kilometer/usage-dependent costs, were broken down to a single parcel. For the calculation of the processing times, time for loading, unloading, delivering, transshipping and parking was considered. Regarding CO_2-emissions, only emissions from energy consumption and thus, driving were considered, by employing respective emission factors.

3 Results

The following section presents the results of the online survey and key insights from the company interviews, followed by an in-depth analysis of combining SEVs with micro-hubs in the CEP industry.

Table 1 Logistics parameters

Parameter	Unit	Van	HDT	Cargo bicycle	Source
Total number of stops/trip	#	87.5	2	16	Own calculation based on interviews
Total numbers of parcels/vehicle	#	175	1,120	32	
Number of parcels/swap bodies per stop	#	2	1	2	Interview data
Available vehicle volume	m^3	12	69.1	2	Own calculation based on interviews [6, 8–11]
Capacity utilization	%	90	90	100	Interview data
Used vehicle volume	m^3	10.8	62.2	2	Own calculation based on interviews
Loading time per parcel/swap body	min	0.5	16	0.5	Own assumption
Loading time per trip	min	87.5	32	16	Own calculation based on interviews
Unloading time per stop (with two parcels/one swap body)	min	3	10	2	[12, 13]
Time loss per stop (parking)	min	1	0	0	Own assumption
Un-/loading time per tour	min	350	52	32	Own calculation based on [14]
Duration of stop between tours	min			16	Own calculation based on Interviews
usable for charging	min			16	Interview data
Duration of warehouse stop	h	14	14	14	
usable for charging	h	12	12	13	
Length of trip	km	80	20.78	12	Own assumption based on interviews
in town	km	66	6.78	12	Own assumption
out of town	km	14	14		
Number of trips per vehicle and day	#	1	4	7	Interview data
Distance between two stops (in town)	km	0.74			Own assumption
Distance to first stop	km	8			
in town	km	1			
out of town	km	7			
Average speed in town	km/h	30	22		Expert knowledge
Average speed daytime out of town	km/h	50			Expert knowledge, own assumption

Table 2 Techno-economic parameters of vehicles

Parameter	Unit	Van	HDT	Cargo bicycle	Source
Number of trips	#/d	1	4	7	Own
Operating life	y	8	10	6	calculations, [15]
Vehicle price	EUR	40,000	280,886	15,000	Following [16–18]
Fuel consumption	l/100 km kWh/ 100 km	11	158	1.6	Following [15, 17]
Fuel price	EUR/l EUR/kWh	1.066	0.184		[19–21]
Maintenance, repair, and tire costs	EUR/km	0.11864	0.3976	900 p.a.	[15, 17, 22, 23]
Tax	EUR/y	211	1,336		[17]
Insurance costs	EUR/y	4,400	7,105	900	[17, 23]
List price charging point	EUR		10,000	2,000	[24, 25]
Operating life infrastructure	y		15	15	[26]
List price batter	EUR			719	[27]
Operating life battery	Y			6	[15]

Table 3 Further techno-economic parameters

Parameter	Unit	Value	Source
Driver wage daytime	EUR/h	25	[28]
Driver wage nighttime	EUR/h	28.8	[16]
Working days per year	d/y	300	[14]
Area of swap body	m^2	18	Own calculation
Rent	EUR/(month*m^2)	11.76	Research on rental portal
Annual rent	EUR/y	2,545	Own calculation
Emission factor WTW diesel	kg CO_2/l	2.964	[16]
Emission factor electricity	kg CO_2/kWh	0.303	

3.1 Results of the Online Survey: Diffusion of, Motives for and Barriers to the Use of SEVs

In a first step, the results from the survey were analyzed to determine the extent to which SEVs are known or are already in use. Based on these findings, the company-specific reasons for or against the use of SEVs were then determined in a second step.

The general knowledge about SEVs in the companies surveyed is evenly split between those who know about SEVs and those who do not. Accordingly, 50% of the respondents have not yet heard of SEVs. 41% of respondents were familiar with two-wheeled cargo bikes. 26% were aware of light electric vehicles, slightly more than the 25% familiar with three- or four-wheeled cargo bikes (see left-hand part of Fig. 2). Once the current state of knowledge about SEVs had been identified, the next step was to ask whether the use of SEVs was generally conceivable from the viewpoint of the companies. For this purpose, all the companies were surveyed, i.e. those already familiar with SEVs and those not familiar with them. The results were that 3% of the companies said they already use SEVs and another 25% can imagine doing so. 71% cannot imagine using SEVs in their company at all (see pie chart in Fig. 2). This means that more than a quarter of all the surveyed companies can imagine using SEVs.

In this context, the question arises about the size of the companies that consider SEVs a potential solution. Are these mainly large companies or is the potential use independent of company size? Looking at the applicability of SEVs by company size as depicted in Fig. 3, it is noticeable that 2.5–5% of all sized companies already use SEVs (although it should be noted here that the absolute figures are very low and range between one and six). In addition, parallel patterns emerge when looking at company size. With the exception of companies with more than 250 employees, 22–27% can imagine using SEVs, while 68–76% cannot in each size group. For companies with more than 250 employees, the use of SEVs appears considerably more feasible. Here, 55% of the companies state that they can imagine using SEVs. Thus, although there were only a few large companies in the sample, they seem to offer high potential for the use of SEVs. All in all 102 companies can imagine using SEVs or are already using them. This willingness seems to exist across all company sizes, although it is particularly marked in large companies (see Fig. 3).

Moreover, this willingness seems also to exist across all sectors. It should be noted that the number of companies that replied is in some sectors quite small. However, with the exception of "Transport and storage" and "Financial and insurance services", companies from all sectors consider SEVs to be applicable in principle (see Fig. 4).

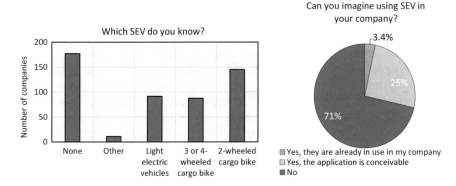

Fig. 2 Knowledge and applicability of SEVs in companies (own representation)

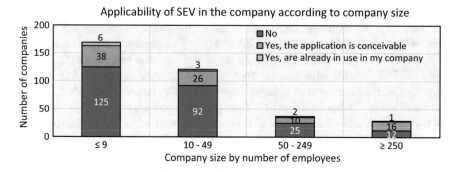

Fig. 3 Applicability of SEVs in companies according to size

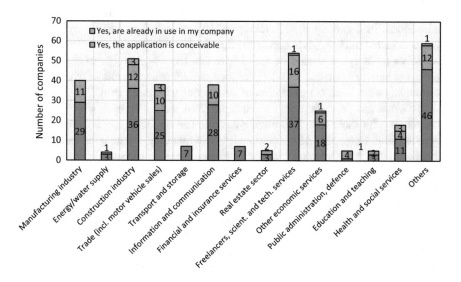

Fig. 4 Applicability of SEVs according to sectors

In this context, it is of interest to find out what advantages companies hope to gain from using SEVs and the 102 companies were questioned about their motives (see Fig. 5). Environmental protection seems to be the most important factor for using SEVs for micro, small and large companies. Reducing fuel costs is also cited as a strong motive. Furthermore, it is noteworthy that micro and small enterprises often have similar motives. In addition, being perceived as innovative is significantly more important for large companies than for other company sizes.

While the company-specific aspects that motivate a company to use SEVs are of interest, it is also essential to know which company-specific factors hinder the use of SEVs. An analysis was therefore conducted using the data from the companies that had previously stated that the use of SEVs was out of the question. Of the 254 respondents who could not imagine using SEVs in their company, 129 said the

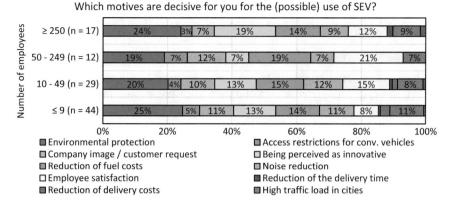

Fig. 5 Motives for the (possible) application of SEVs

reason is that they have no problems with their existing logistics that SEVs could solve. Seventy-seven mentioned that the range of SEVs was considered too limited, followed by insufficient transport volume and too small payload capacity with 64 and 63 mentions, respectively. Eighteen responses mentioned the lack of micro-depots as hindering the use of SEVs. It should be noted, however, that the lack of micro-depots may not be relevant in every economic sector. Examining the obstacles to use SEVs according to company size, it is noticeable that companies with up to 249 employees do not name logistics problems as the main reason. In the case of large companies, the aspect of insufficient range is mentioned most frequently. Figure 6 illustrates the results depending on the company size.

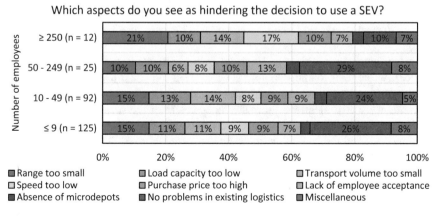

Fig. 6 Barriers to the use of SEVs

3.2 Supplementary Findings from Interviews with Companies from Potentially SEV-Relevant Sectors

In order to gain a better understanding of the results of the online survey and why SEVs are considered or not considered by companies in a particular sector, supportive in-depth interviews were conducted with companies from sectors where SEVs might be a relevant option. A brief summary of the most important results of the interviews is presented below.

The statements of the interviewed companies showed that especially nursing services, CEP services, pharmacies, and internal factory traffic seem to be suitable for SEVs. The installation and chimney-sweeping sectors could integrate and use SEVs to some extent in their everyday work. Delivery services are, in principle, also suitable for SEVs in urban areas. Delivery speed is of enormous importance here, especially in the gastronomic sector, and SEVs are not always the fastest option.

It must be emphasized that it is difficult to make a general statement about the suitability of SEVs for an entire sector, as the requirements of companies in the same sector can already be very diverse and therefore a company-specific evaluation should always be made. For example, in the painting business, there are quite different requirements for the range and cargo weight of vehicles. Furthermore, interview partners from the chimney-sweep sector stated that the structure of their working districts varies widely. For example, one district has a diameter of 5.5 km and can be easily covered on foot. In contrast, other districts, which include single-family houses, for example, can be up to 17 km wide and would be suitable for SEVs.

An application of SEVs in the field of nursing services is quite conceivable. The advantages mentioned were more efficient and faster trips as well as reduced time needed to find a parking lot. Another major advantage is that employees without a category B driver's license could also use them if necessary. Using SEVs could therefore enlarge the pool of potential job applicants, as possession of a category B driving license would no longer be essential. This may also be an issue for other sectors.

In the installation sector, SEVs seem to be particularly suitable for customer services, while their loading volume is considered too small for cleaning services. They cannot transport ladders and larger machines, for example. In addition, other obstacles were also mentioned that are independent of the sector, such as the fact that company cars are sometimes also used privately and SEVs do not meet these requirements, and the lack of charging infrastructure at home. The potential lack of weather protection and lower performance in winter were also criticized. It should be mentioned that the results of the interviews are to be regarded as exemplary and not representative due to the different framework conditions of companies, even within the same sector.

3.3 Results from the In-Depth Interviews and Quantitative Analyses: Acceptance and Impact of SEVs for Micro-Hubs in the CEP Industry

Using SEVs, such as three- or four-wheeled cargo bicycles, in commercial transportation is being increasingly discussed in combination with innovative city logistics concepts. In the logistics literature, one of the most promising solutions is to use SEVs for last-mile distribution from urban micro-hubs or depots [29, 30]. We therefore shed more light on the acceptance of combining electric cargo bicycles with urban micro-hubs and their economic and environmental impacts.

We interviewed CEP LSPs (5), logistics departments of wholesalers for groceries, paintings or other goods (3), transport companies (2), a distributor of a bakery chain (1), a distributor of a trading company (1) and a distributor of newspapers (1). The expert interviews revealed that six out of the thirteen interviewed LSPs would be willing to implement micro-hubs with cargo bicycles. Another LSP would probably implement that concept, while six LSPs would not implement it all (see Fig. 7).

The main reasons given for rejecting the concept of cargo bicycles with micro-hubs were very diverse. Most of the interviewees mentioned too low transportation capacities or the weight restrictions of cargo bicycles as the main reason. These LSPs tend to transport large and/or heavy goods, which are not suitable for cargo bicycles or SEVs in general. Another reason was missing cooling capabilities and difficulties with adhering to hygiene standards. Large quantities of goods that cannot be distributed efficiently with that concept and increased transportation lead-times due to transshipping at the micro-hubs were further reasons for rejecting this concept. Finally, the lack of profitability and a limited number of personnel or difficulties in recruiting new drivers were mentioned. In contrast, other companies responded that they had already implemented cargo bicycles with micro-hubs, at least in first pilot projects. Advantages were mentioned such as improved transportation cost efficiencies for the last mile and circumventing urban access regulations, especially in the morning, in the evening or during the night.

The logistics departments of wholesalers for painting or groceries were mostly reluctant to implement cargo-bicycles with micro-hubs, as were bakeries and pharmacies. Transport companies and general cargo carriers also rejected the

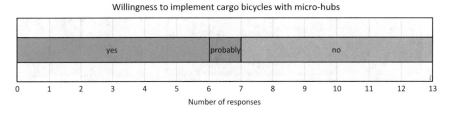

Fig. 7 Willingness of LSPs to implement cargo bicycles in combination with micro-hubs

concept, mostly because their transported goods are too big and too heavy. Finally, one CEP LSP also rejected the implementation of micro-hubs. Distributors of newspapers, however, as well as other CEPs were willing to implement cargo bicycles with micro-hubs. One distributor of textiles could imagine using cargo bicycles with micro-hubs to distribute textiles ordered online to end-customers, while micro-hubs would not be an option for deliveries to retail stores.

In summary, acceptance of SEVs depends heavily on the logistics structures and general framework conditions, as well as on the type of transported goods. This is very industry-specific and the results show that the highest potential for cargo bicycles with micro-hubs is in the distribution of smaller goods, mainly to end-customers, which primarily concerns the CEP industry.

Based on these results, and considering the high acceptance of this concept among CEP LSPs, a quantitative analysis of the economic and environmental impacts was then carried out.

Figure 8 shows the results of calculating the TCO, CO_2 emissions and processing time. Shifting from the status quo to cargo bicycles with micro-hubs would decrease transportation costs per parcel by 25%. The main reason for this is that one cargo bicycle can actually replace one delivery van and cargo bicycles are much cheaper in terms of investments, as well as variable costs. We assumed that one van distributes 175 parcels on average on a standard tour lasting 9.7 h. The loading capacity of a cargo bicycle was assumed to be 32 parcels and its distribution tours only take around 1.3 h. In order to distribute the same number of parcels as a van, six bike tours are required, adding up to 9.4 h and including transshipping time at the micro-hub.

Cargo bicycles with micro-hubs can reduce the CO_2 emissions per parcel by almost 80% (see Fig. 8). It has to be mentioned here that we only included the emissions from driving and excluded the emissions from the production of the

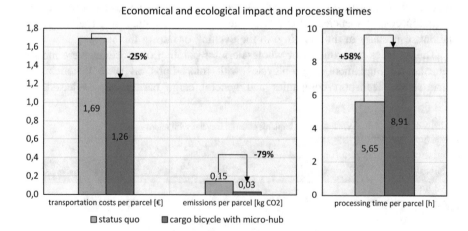

Fig. 8 Results of economical, ecological, and processing time calculations

vehicles. Nevertheless, this represents a huge reduction. This is because electric drivetrains are much more energy-efficient than conventional internal combustion engines and electricity has a much lower CO_2 emission factor than diesel. Reducing the emissions to zero would only be possible using 100% renewable electricity, which is currently not the case (calculations are based on Germany's power mix). Finally, average processing times would increase by almost 60%. The processing time starts when a parcel leaves the regional distribution center and ends when it is delivered to the recipient. Thus, for the status quo, the parcel in the van is more or less directly on its way or at least on its final distribution tour to the recipient. For the micro-hub case, however, we assume that the micro-depots have to be distributed by the HDT (2 per HDT-tour), which takes extra time, as does trans-shipping at the micro-hub and only then can the cargo bicycle tour start from the micro-hub.

To sum up, from an economic and environmental perspective, electric cargo bicycles in combination with micro-hubs offer major benefits. Costs can be reduced by 25%, while CO_2 emissions can be reduced by 79%. The only drawback is the higher processing time involved, which increases by 58%.

4 Summary and Conclusions

SEVs can contribute to more sustainable commercial transportation due to their reduced size, lower noise emissions, and lower CO_2 emissions. However, so far, only half of the companies surveyed know about them. Nevertheless, a quarter of these companies can imagine using SEVs, so there is a large user potential for SEVs in commercial transport. This potential is found in all sizes of companies, but especially in large ones. The main reasons for potentially using SEVs are not primarily monetary. Environmental protection and employee satisfaction are ranked first and second here. Thus, policy makers could use this intrinsic motivation to promote the diffusion of SEVs and thus reduce THG emissions. The reasons given for not using an SEV include no logistical advantage for the company and construction-related criteria such as transport weight, range or load volume. In order to activate the as yet unexploited potential, it is therefore advisable for SEV sellers to further inform companies about the existence and advantages of SEVs. In addition, the trial use of SEVs could offer the opportunity to show potential areas of application in companies that have not yet considered them.

The acceptance of electric cargo bikes in combination with innovative city logistics concepts, such as micro-hubs, is very high. Almost half of the interviewed LSPs would be willing to implement such a concept. However, acceptance depends to a large extent on the characteristics of the transported goods and thus, on the type and sector of the LSP. Bearing in mind, that electric cargo bikes would not be able to transport the bulk goods of many of the interviewees, the acceptance is high. The CEP industry, in particular, shows high acceptance of this concept, which offers economic and environmental benefits, but increased average processing times. The necessity of fast deliveries, however, is arguable. Consequently, for LSPs, electric cargo bicycles in a micro-hub concept represent an economically and environmentally beneficial solution with the potential to decrease urban traffic and make our cities greener.

Acknowledgements This contribution was written in the framework of the Profilregion Mobilitätssysteme Karlsruhe, which is funded by the State Ministry of Economic Affairs, Labour and Housing in Baden-Württemberg, and as a national High Performance Center by the Fraunhofer-Gesellschaft.

References

1. Fu, J., Jenelius, E.: Transport efficiency of off-peak urban goods deliveries: a Stockholm pilot study case studies on transport policy **6**, 156–66 (2018)
2. Kampker, A., Deutskens, C., Maue, A., Hollah, A.: Elektromobile Logistik. In: Deckert, D. (ed.) CSR und Logistik. Spannungsfelder Green Logistics und City-Logistik, 1st edn, pp. 293–308. Springer Gabler (Management Series Corporate Social Responsibility), Berlin, Heidelberg (2016)
3. EEA: National emissions reported to the convention on long-range transboundary air pollution (LRTAP Convention). European Environment Agency, Copenhagen (2018)
4. Holguín-Veras, J., et al.: Direct impacts of off-hour deliveries on urban freight emissions. Transp. Res. Part D: Transp. Environ. **61**, 84–103 (2018)
5. Prümm, D., Kauschke, P., Peiseler, H.: Aufbruch auf der letzten Meile. PricewaterhouseCoopers GmbH Wirtschaftsprüfungsgesellschaft, Neue Wege für die Städtische Logistik (2017)
6. Rumscheidt, S.: Die letzte Meile als Herausforderung für den Handel - ifo Schnelldienst **72** (1), 46–49 (2019)
7. Brost, M., Ewert, A., Schmid, S., Eisenmann, C., Gruber, J., Klauenberg, J.: Elektrische Klein-und Leichtfahrzeuge: Chancen und Potenziale für Baden-Württemberg. Studie im Auftrag der e-mobil BW (2019)
8. BIEK: Innovationen auf der letzten Meile: Kurier-, Express- Paketdienste (2017)
9. Graf, H.-W.: Optimierung des Wechselbrückentransports - ein Spezialfall der Tourenplanung Große Netze der Logistik: Die Ergebnisse des Sonderforschungsbereichs 559 ed P Buchholz and U Clausen (Berlin: Springer), pp. 101–28 (2009)
10. RYTLE GmbH: MOVR: Technische Daten, (2019). https://rytle.de/movr/. Last Accessed 2 Oct 2019
11. KRONE GmbH & Co. KG: Dry Box: Typ: WK7,3 STG (2018)
12. Holguín-Veras, J., et al.: The New York City off-hour delivery program: a business and community-friendly sustainability program. Interfaces **48**, 70–86 (2018)
13. Bogdanski, R.: Quantitative Untersuchung der konsolidierten Zustellung auf der letzten Meile (2019)
14. Arndt, W.-H.: Trends im urbanen Lieferverkehr Lieferkonzepte in Quartieren - die letzte Meile nachhaltig gestalten: Lösungen mit Lastenrädern, Cargo Cruisern und Mikro-Hubs ed W-H Arndt and T Klein (Berlin Difu), pp. 5–9 (2018)
15. Moll, C.: Nachhaltige Dienstleistungsinnovationen in der Logistik: Ein Ansatz zur Entwicklung von Entscheidungsmodellen. Springer Fachmedien Wiesbaden, Wiesbaden (2019)
16. Rosenberger, T., et al.: Der Lastauto Omnibus Katalog 2018. EuroTransportMedia Verlags-und Veranstaltungs-GmbH, Stuttgart (2017)
17. Mercedes-Benz, A.G.: Der neue Sprinter - Kastenwagen Tourer Fahrgestelle: Preisliste. Gültig ab 8. Februar 2018. Daimler AG (2018)

18. Wietschel, M., et al.: Machbarkeitsstudie zur Ermittlung der Potentiale des Hybrid-Oberleitungs-Lkw. Fraunhofer ISI, Karlsruhe (2017)
19. IEA: World Energy Outlook 2018, International Energy Agency (2018)
20. Bundesverband der Energie und Wasserwirtschaft e. V., BDEW (2019) BDEW-Strompreisanalyse Juli 2019: Haushalte und Industrie, Berlin
21. Gerbert, P., et al.: Klimapfade für Deutschland, BCG and Prognos AG (2018)
22. Onat, N.C., Kucukvar, M., Tatari, O.: Conventional, hybrid, plug-in hybrid or electric vehicles? State-based comparative carbon and energy footprint analysis in the United States. Appl. Energy **150**, 36–49 (2015)
23. citkar GmbH: Loadster: Der Loadster All-Inclusive, (2019). https://loadster.org/produkt/loadster-all-inclusive/. Last Accessed 6 Nov 2019
24. Hacker, F., Waldenfels, R., Mottschall, M.: Wirtschaftlichkeit von Elektromobilität in gewerblichen Anwendungen: Betrachtung von Gesamtnutzungskosten, ökonomischen Potenzialen und möglicher CO2-Minderung, Öko-Instiut e.V (2015)
25. NPE: Ladeinfrastruktur für Elektrofahrzeuge in Deutschland: Statusbericht und Handlungsempfehlungen 2015, Berlin (2015)
26. Urban-e GmbH & Co.KG: Urban-e: Cargo eBikes for professionals!: KEP10, (2019). https://www.urban-e.eu/. Last Accessed 2 Oct 2019
27. Schroeder, A., Traber, T.: The economics of fast charging infrastructure for electric vehicles. Energy Policy **43**, 136–144 (2012)
28. Holguín-Veras, J., et al.: Fostering unassisted off-hour deliveries: the role of incentives. Transp. Res. Part A: Policy Pract. **102**, 172–187 (2017)
29. Bienzeisler, B., Bauer, M., Mauch, L.: Screening City-Logistik. Europaweites Screening aktueller City-Logistik-Konzepte, Fraunhofer IAO (2018)
30. Dolati Neghabadi, P., Evrard Samuel, K., Espinouse, M.-L.: Systematic literature review on city logistics: overview, classification and analysis. Int. J. Prod. Res. **57**(3), 865–887 (2019)

An Energy Efficiency Comparison of Electric Vehicles for Rural–Urban Logistics

Andreas Daberkow, Stephan Groß, Christopher Fritscher, and Stefan Barth

Abstract In many small and medium-sized businesses in rural–urban areas, delivery services to and from customers, suppliers, and distributed locations are required regularly. In contrast to purely urban commercial centres, the distances here are larger. The aim of this paper is to identify opportunities for substituting combustion-engine logistics with lightweight electric commercial vehicles and the limitations thereto, describing an energy efficiency comparison and improvement process for a defined logistics application. Thus, the area of Heilbronn-Franconia and its transport conditions are presented as examples to compare the use case to standard driving cycles. Then the logistic requirements of Heilbronn UAS (*University of Applied Science*) locations and the available vehicles as well as further electric vehicle options are depicted. Options are discussed for the additional external payload in search of transport volume optimisation without increasing the vehicle floor space. To this end, simulation models are developed for the aerodynamic examination of the enlarged vehicle body and for determining energy consumption. Consumption and range calculation lead to vehicle concept recommendations. These research activities can contribute to the transformation of commercial electro mobility in rural and urban areas in many parts of Germany and Europe.

Keywords Small electric commercial vehicle · Rural–urban logistics · Computational fluid dynamics · Transport volume optimisation

1 Introduction

When considering future transportation options, heavy-duty vehicles and their alternative drives come to mind, although a commercial approach is lacking. Both opportunities and challenges seem immense, although available technologies enable

A. Daberkow (✉) · S. Groß · C. Fritscher · S. Barth
Heilbronn UAS, Max-Planck-Straße 39, 74081 Heilbronn, Germany
e-mail: andreas.daberkow@hs-heilbronn.de

© The Author(s) 2021
A. Ewert et al. (eds.), *Small Electric Vehicles*,
https://doi.org/10.1007/978-3-030-65843-4_7

change towards more sustainable transport for the vast majority (76%, [1]) of commercial vehicles in Europe. The sub-3500 kg-N1-class of commercial vehicles defined by EU regulations [2] does not cover long-distance freight shipping and only handles local to rural individual end-customer supply. This class with payloads comparable to passenger cars still lacks alternative powertrains (<2% [1]), despite the continuously rising share of electric passenger car production (12% of the German Car Industry by April 2020, [3]). Road logistics powertrain electrification has slowly been growing [4]; particularly on shorter tracks, electric commercial vehicles have come into mass use [5], due to stricter emission limits. Moreover, battery electric vehicles (BEV) hold the potential of logistics cost reduction [6].

The possibility of concept transfer from passenger BEVs to rural transporters and application of the ecological imperatives begs the question of how to select the most appropriate vehicle for the particular application, which is answered in the following paragraphs.

1.1 Developments in Rural BEV Application

In the field of inner-city delivery traffic with light commercial vehicles [7, 8], as also for heavy commercial vehicles on the "last mile" in urban distribution traffic [9, 10], there have already been numerous developments, investigations, and studies. Delivery services are increasingly employing electric vehicles in cities [11, 12].

The range of electrically powered commercial vehicles limits their use initially to urban areas. Hardly any scientific publications have investigated the potential of commercial BEVs in rural–urban areas. The "*eMiniVanH*" project established by the Ministry of Economic Affairs Baden-Wuerttemberg aims to fill this gap.

Important features of vehicles used in rural–urban areas are longer distances and higher driving speeds. Daberkow and Häussler [13] describe the usability of light electric passenger cars in this rural–urban area, as a first investigation. They range in variety from specially developed research vehicles [14, 15] to models already available on the market from well-known vehicle manufacturers. A collection of some small electrical vehicles is given by Brost et al. [16].

The application of such vehicles in courier and parcel delivery services creates a demand for a daily range of 30–800 km [17, p. 171]. By limiting these driving profiles to more task-related parcel services, the range requirement shrinks to 30–360 km, with average speeds up to 60 kmh^{-1}, soon to be covered by common BEVs.

N1 light-duty BEVs try to enter a most competitive market segment, which eliminated several small companies and small series of large OEMs. Therefore, the following legally highway-suitable vehicles are all considered, and chosen as representative types for further discussion because they differ significantly in size and load volume: the *Renault Kangoo Zero Emission* (2013–2017), an electrified high roof station wagon; the *Streetscooter Work Box* (2015–2020), solely battery electric, developed for *Deutsche Post AG*; the *Volkswagen e-up! load-up!* (2013–

2016), an electric light-duty variant of a mini car; and finally, the *Volkswagen e-up!* (since 2020), with its extended range facelift. From these types, a vehicle is chosen matching the use case, which can replace a combustion vehicle most efficiently.

1.2 Facility Test Environment in Heilbronn-Franconia Region

The "eMiniVanH" project deals exemplarily with freight traffic between the Heilbronn UAS locations. The state of Baden-Wuerttemberg lies along the French border and is located in the southwest of Germany. The Heilbronn-Franconia region, see Fig. 1, covers an area of 4765 km^2 with a population of roughly 0.9 million, and its administrative seat is Heilbronn (population 130,000), see [18].

Heilbronn-Franconia is an important economic region. Large manufacturers like *AUDI AG* as well as large suppliers like *Robert Bosch GmbH* contribute to the economic wealth of the region.

Individual mobility and public transportation are key aspects of the region. UAS has purchased and operates a *VW e-up! load-up!* model as representative of a small electric commercial vehicle, see Fig. 2 left and middle.

This special vehicle has a continuous cargo area instead of the rear row of seats. With 60 kW drive power and an installed battery capacity of 18.7 kWh, this compact vehicle (length 3540 mm according to VW AG [20]) is eminently suited to urban as well as rural areas. The initial tests were made for parcel and mail transport substitution (see Fig. 2 on the right). As the standard freight consists of a few post

Fig. 1 Heilbronn landscape [19] and rural–urban location of Heilbronn-Franconia region [18]

Fig. 2 Volkswagen *e-up! load-up!* with cargo compartment (left and middle) and a typical example of parcel and mail transport with an internal combustion-engine-powered transporter (right)

boxes, replacing the combustion-engine-powered transporter is easily possible. About 960 kg of the payload is to be transported, per week. However, further expansion of the loading volume is desirable for additional applications, and the effects on the range must be investigated.

2   Digital Prototypes and Simulated Driving Cycles

Prior to prototype manufacturing and road test execution, a preceding digital part development supported by simulations must prepare design decisions. As with many car and truck bodybuilders, digital data of the base vehicle, for example, CAD-3D or Digital Mock-Up (DMU) data for the *VW e-up! load-up!* are not available. The following Sect. 2.1 describes the reverse engineering of a digital prototype for further investigations. In Sect. 2.2, this DMU is assessed aerodynamically, as the air drag is mainly of relevance for energy consumption simulations. Section 2.3 describes energy consumption simulations for the use case, a facility management trip between all four locations, and other scenarios comparing several competing vehicle variants.

2.1   Creating a Digital Mock-up

The DMU also provides the opportunity to design extra volumes for transport and load carrier fixation systems for the specific case. The digital representation does not require all details and parts of the vehicle. Only exterior surfaces and interior geometry of the cargo bay are of relevance. Manual Laser imaging, detection and ranging (LIDAR) scanning produces STL-Files of the payload compartment and the exterior surfaces, as shown in Fig. 3. Some errors occur while matching several scans together automatically. Redesigned post and pharma boxes complete the DMU.

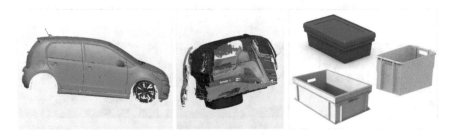

Fig. 3 LIDAR-Scan of the *VW e-up! load-up!*; left: exterior; centre: cargo compartment; right: pharmaceutical cargo containers, "*Postbehälter Typ 2*" and VDA/Euronorm

2.2 CFD Based Roof Extension Development

The roof of the *VW e-up! load-up!* is also used for the generation of storage space, in addition to the interior space. The design of potential roof box variants is based on computational fluid dynamics (CFD) simulations. The models of the *VW e-up! load-up!* and its roof extension variants are shown in Fig. 4. Four post boxes stored in either container extend the storage volume by 100 l, as the roof load is restricted to 50 kg, equalling four times the mass capacity of a post box.

The Reynolds-Averaged Navier-Stokes K-Ω model and a steady-coupled implicit flow solver were used to simulate the turbulent flow of the incompressible air with 20 ms^{-1} (72 kmh^{-1}) and 35 ms^{-1} (126 kmh^{-1}) for comparison. As the model contains ten prism layers geometrically growing with a growth factor of 1.73 over 8 mm total thickness, all the wall-y+ values lie below 3 as required by the applied turbulence model. The wheels rotate at matching angular velocity, their separate rim mesh region consisting of a moving reference frame. The vehicle geometry is simplified by a closed radiator grille and a smooth vehicle undertray and neglects suspension components. Tire treads, mirrors, and wheel front flicks are considered. Exploiting symmetry properties reduces the cell count to 16 million by using a half model in an open road setup [21].

Different roof extension designs are compared to the scanned reference model using the CFD results drag coefficient c_d and the normal area in the driving direction A_x. The objective is a minimized additional air resistance for the predefined load volume gain.

2.3 Driving Cycles for the Rural–Urban Use Case

Today, a vehicle's energy consumption is compared utilising standardised test procedures, the New European Driving Cycle (NEDC) and the Worldwide Harmonised Light Vehicles Test Procedure Class 3b (WLTP) driving cycles predefined by UN-laws UN ECE/324 and UN GTR15. The WLTP Class 3b driving cycle provides a good basis for vehicle comparison in a rural–urban use case.

These predefined cycles may not necessarily represent the specific requirements of arbitrary delivery services in rural–urban areas. Here, a further unique Use Case

Fig. 4 e-up! *load-up!*, removable roof box (middle), fixed high roof compartment (right)

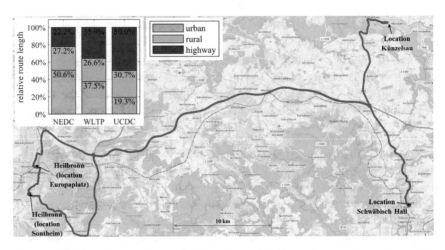

Fig. 5 Driving route between the UAS Heilbronn campus [own illustration with map material from © 2020 GeoCzech, Inc.] and chart with street and traffic type characteristics (top left)

Driving Cycle (UCDC) for the Heilbronn UAS testbed completes the assessment as a third cycle. This route as shown in Fig. 5 connects the different campuses of the UAS.

The UAS has two campuses in Heilbronn, one in Künzelsau and the other in Schwäbisch Hall. The route with 145 km total length contains city traffic, rural roads, and highways. Its sections represent a real use case with street and traffic types as in Table 1.

NEDC consists of two parts (urban and non-urban), and WLTP Class 3b distinguishes four different speed sections. Figure 5 shows that the UCDC lacks city tracks but has a larger share of highway track length compared to the WLTP. This partial route with a top speed of 100 kmh^{-1} contributes to a smoother but faster cycle on average with an average absolute acceleration $\overline{|a|} = 0.230$ ms^{-2} (WLTP: $\overline{|a|} = 0.358$ ms^{-2}) and an average velocity $\bar{v} = 64.8$ kmh^{-1} (WLTP: $\bar{v} = 46.5$ kmh^{-1}). Consumption and range calculation determine transferability of WLTP results for the UCDC.

Table 1 Characteristics of different street and traffic types [22]

Street type	Characteristics
City streets with urban traffic	Driving speed up to 60 kmh^{-1}, frequent stop-and-go, intermittent acceleration necessary
Country roads	Driving speed between 60 and 90 kmh^{-1}, no stops, certain acceleration necessary
Highway	Driving speed up to 130 kmh^{-1}, constant driving, hardly any acceleration necessary

2.4 Simulation Model for Vehicle Drive Cycles

To compare the energy consumption stated at the accumulators of different vehicles, a *MATLAB®* program evaluates the velocity profile of the three driving cycles. Dry mass, payload, acceleration, and velocity contribute to the driving resistance forces' drag, tire friction, and inertia and their corresponding powers [23]. The non-constant altitudes of the UCDC are included. Due to restricted public access, several parameters were estimated and used equally for all assessed vehicles, as Table 2 shows.

The *VW e-up! load-up!* (2013) has a payload capacity of 286 kg, which shall be the payload in the presented use case. Technical data of each vehicle provides their dry mass and dimensions, but no information about drag coefficients' projected frontal area is published. For an engineering estimation, a cross-sectional CAD-sketch delivers well-approximated values. Drag coefficients' estimations are listed in Table 3.

In addition, technical data deliver values for battery capacity used for range calculation and at least one driving-cycle-based consumption value. The *VW e-up! load-up!* OEM data shows 11.7 kWh/100 km NEDC energy consumption, whereas the simulation without payload shows 11.8 kWh/100 km. The *Kangoo Z.E.* OEM data shows 15.2 kWh/100 km NEDC energy consumption, while the simulation of the empty vehicle shows 15.1 kWh/100 km. These sufficiently matching results qualify the simulation very well for further concept comparisons and thus indicate verifiable results.

Table 2 Substitute parameters used equally for all assessed vehicles

Parameter	Abb.	Value	Source
Tire friction coefficient	f_R	$0.01 + \frac{v}{10^4}\frac{s}{m} + \frac{v^4}{2\cdot10^7}\frac{s^4}{m^4}$	[23, p. 50]
Powertrain efficiency coefficient	η	0.78	[24, p. 124]
Recuperation efficiency coefficient	η_{recu}	$0.741 = 0.95\eta$	[25, p. 19]
Mass surcharge factor for the moment of inertia	k	1.25	[26, p. 82]

Table 3 Aerodynamic parameters of compared vehicles

Vehicle	A_x (m^2)	c_d	Source
VW e-up! load-up! (2013)	2.07	0.311	LIDAR Scan and CFD
Renault Kangoo Z.E.	2.5	0.35	Estimation [27, pp. 66 + 643]
Streetscooter Work Box	3.5	0.45	Estimation [27, p. 643]
VW e-up! (2020)	2.1	0.30	Estimation based on 2013 model

3   Result Evaluation for Designs and Energy Consumption

The first section of this chapter summarises the space gained by enlarging the interior and the roof box. Section 3.2 describes the conceptual decision for the roof box, determined by CFD simulations. Based on this, Sects. 3.3 and 3.4 comprise the simulation results with the consequences for the different vehicle types.

3.1   Enlargement of the Interior Space

There are some design options to enlarge the interior loading capacity to individual requirements [28]. All-purpose solutions or individual custom-made designs are offered by various manufacturers [29].

Especially with small vehicles, optimisation of the already limited loading volume is of critical importance. The simplest way to optimise the loading opportunities is to enlarge the loading floor to the front area by removing the front passenger seat. Furthermore, it must be ensured that the driver's view is not inadmissibly restricted and that the driver is not endangered by the payload [30]. For the universal requirements of the load compartment, a flat loading platform is suitable.

3.2   CFD Simulation Results

The velocity profiles of the simulated flows are shown in Fig. 6.

The acute angle at the beginning of the removable roof box results in a relatively low stagnation point (1). A high loss of velocity occurs in the gap between the roof box and the vehicle roof, which negatively affects the calculated c_d-value (2). Based on the absence of any space between the car roof and the high roof compartment,

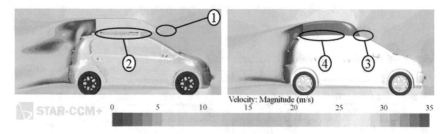

Fig. 6 Simulated flow velocity profiles for roof concepts from Fig. 5. Driving route between the UAS Heilbronn campus [own illustration with map material from © 2020 GeoCzech, Inc.] and chart with street and traffic type characteristics (top left) Fig. 4

Table 4 Comparison of the calculated values

	VW e-up! load-up!	+ roof box	+ high roof variant
Drag coefficient c_d	0.311	0.44	0.343
Reference area A_x (m^2)	2.07	2.30	2.31
$c_d{\cdot}A_x$ (m^2)	0.646	1.014	0.791
Drag force (20 ms^{-1}) (N)	**152**	**239**	**186**

the drag forces in this area are significantly lower than for the removable roof box (4). In addition, the high roof is in contact with the vehicle body, thereby making for optimal deflection at the beginning of the high roof. Moreover, no direct stagnation point is created at the top of the high roof (3). Table 4 compares simulation results for the roof extensions to the *VW e-up! load-up!* model equipped as standard.

The published drag coefficient of 0.308 [20] for *the VW e-up! load-up!* is slightly below the CFD simulation result of 0.311 (see Table 4). In comparison, the high roof compartment delivers far better results than the removable roof box. This happens because the high roof variant has no gap between vehicle body and high roof and therefore does not lead to unfavourable flow conditions.

3.3 Results of the Simulated Drive Cycles

Despite the differences between the real-driven UCDC and the standardised test-bench cycle WLTP, both lead to the same consumption, as Fig. 5 shows, differing less than 2%. Despite the significant differences outlined in Sect. 2.3, the WLTP represents this use case adequately. The NEDC results in 13–21% less electric energy usage, depending on the assessed vehicle. In conclusion, the simulated WLTP provides an appropriate prognosis for small and light commercial BEV energy consumption.

The *VW e-up! load-up!* stands out regarding consumption, even fully loaded. Additional extensions like the examined roof compartment increase aerodynamic resistance to such an extent that a high roof station wagon type becomes the recommended vehicle concept, as it offers around 400% more cargo space with approximately the same air resistance. Neither consumption nor range qualifies the *Streetscooter* or comparable vehicle types for operation in this use case, as their design for solely urban terrain is reflected in aerodynamic weakness, as Fig. 7 shows.

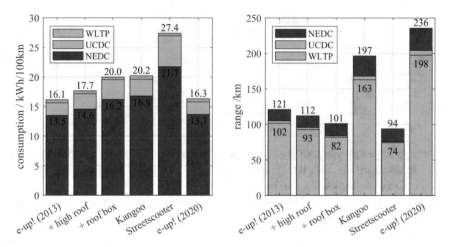

Fig. 7 Energy consumption and range calculation results

4 Conclusion

The standard variant of the *VW e-up! load-up!* vehicle type can carry 12 post containers, resulting in 300 l cargo volume. Due to the proposed interior design change of the *VW e-up! load-up!*, 20 post containers of 500 l in total are available. This is 67% more cargo volume than the reference model. This approach goes beyond solutions with roof extensions, due to the absence of aerodynamic deterioration. The *VW e-up! load-up!* high roof variant together with the new interior design, see Fig. 6, is designed for four additional post containers with a total of 600 l. Compared to the standard variant, this yields 100% more containers, although it leads to an increase in energy consumption by 20%. The range with a high roof thus decreases from 102 to 82 km, see Fig. 7. Before adding roof storage to a light-duty mini car, deciding on the *Kangoo* is thus more energetically reasonable.

The developed cargo load concept and the energy consumption investigations show that the *VW e-up! load-up!* vehicle types are a good option close to small and light electric commercial vehicle concepts for the rural–urban region with larger distances. Even at higher speeds, acceptable distances and payloads can be covered without stopping for charging. Thus, the *VW e-up!* (2020) including an enlarged, interior payload compartment becomes the ideal choice for the presented use case.

References

1. Adolf, J., Balzer, C., Haase, F., Lenz, B., Lischke, A., Knitschky, G.: Shell commercial vehicle study. Shell Deutschland Oil GmbH, Hamburg. https://bit.ly/3ej0M6B (2016)
2. N.N.: Commission Regulation (EU) No. 678/2011 of 2011/07/14
3. VDA-Homepage. https://bit.ly/3cYWUq4. Accessed 2020/04/26
4. Witzig, J., Wenger, M., Januschevski, R.: Elektrischer Zentralantrieb für Nutzfahrzeuge. In: MTZ—Motortechnische Zeitschrift, vol. 10 (2018)
5. Burkert, A.: Elektroantrieb im Nutzfahrzeug. In: MTZ—Motortechnische Zeitschrift, vol. 06 (2019)
6. Kampker, A., Deutskens, C., Müller, P., Müller, T.: Reduzierung der Gesamtbetriebskosten durch den Einsatz von Elektrofahrzeugen. In: ATZ—Automobiltechnische Zeitschrift, vol. 03 (2015)
7. Gumpoltsberger, G., Pollmeyer, S., Neu, A., Hirzmann, G.: Plattform für urbane und automatisierte Elektrofahrzeuge. In: ATZ—Automobiltechnische Zeitschrift, vol. 03 (2017)
8. Höfer, A., Esl, E., Türk, D.-A., Hüttinger, V.: Innovative Fahrzeugkonzepte für Shanghai letzte Meile. In: ATZ—Automobiltechnische Zeitschrift, vol. 06 (2015)
9. Zellinger, M., Wohlfarth, E.: Lokal emissionsfreier und leiser Güterverkehr mit dem Mercedes-Benz eActros. In: MTZ—Motortechnische Zeitschrift, vol. 06 (2018)
10. Schäfer, P.: Volvo Trucks stellt zwei schwere Elektro-Lkw vor. In: Springer-Professionals. https://bit.ly/2KK6UHG. Accessed 2020/04/26
11. Schlott, S.: Entwicklungspfade zum CO2-neutralen Güterverkehr. In: MTZ—Motortechnische Zeitschrift, vol. 05 (2020)
12. Schäfer, P.: UPS bestellt 10.000 Elektro-Transporter von Arrival. In: Springer Professionals. https://bit.ly/3aTpyHK. Accessed 2020/04/26
13. Daberkow, A., Häussler, S.: Electric car operation in mixed urban-regional areas. In: 13th EAEC FISITA conference, 13–16th June, Valencia (2011)
14. Pautzke, F., Schäfer, C., Rischel, W., Zöllner, H., Woeste, G.: Zweckgerichtete Entwicklung eines elektro-Kleintransporter. In: ATZ—Automobiltechnische Zeitschrift, vol. 03 (2011)
15. Lesemann, M., Welfers, T., Mohrmann, B., Eckstein, L.: Konzeption und Aufbau eines elektrischen Lieferfahrzeugs. In: ATZ—Automobiltechnische Zeitschrift, vol. 09 (2014)
16. Brost, M., Ewert, A., Schmid, S., Eisenmann, C., Gruber, J., Klauenberg, J., Stieler, S.: Elektrische Klein- und Leichtfahrzeuge—Chancen und Potenziale für Baden-Württemberg
17. Ludanek, H.: Fahrzeuganforderungen bei leichten Nutzfahrzeugen für den inner- und außerstädtischen Lieferverkehr. In: Karosseriebautage Hamburg, Springer-Vieweg, Wiesbaden (2017)
18. Lage der Region Heilbronn-Franken in Deutschland. By TUBS—Own work, CC BY-SA 3.0. https://commons.wikimedia.org/w/index.php?curid=6334766. Accessed 2020/08/21
19. Heilbronn and surroundings seen from the Michaelsberg. By K. Jähne—Own work, CC BY-SA 3.0. https://commons.wikimedia.org/w/index.php?curid=7992303. Accessed 2020/08/21
20. N.N.: Selbststudienprogramm SSP Der e-up! Volkswagen, Wolfsburg (2014)
21. External Aerodynamics with STAR-CCM + Best Practice Guidelines (v2019.02), Siemens AG (2019)
22. Commission Regulation (EU) No. 427/2016, Real Driving Emissions (RDE)
23. Braess, H.-H., Seiffert, U.: Vieweg Handbuch Kraftfahrzeugtechnik. Springer Vieweg, Wiesbaden (2013)
24. Grunditz, E.A.: Design and assessment of battery electric vehicle powertrain, with respect to performance, energy consumption and electric motor thermal capability. Thesis for the degree of Doctor of Philosophy, Chalmers University of Technology, Göteborg, Sweden 2016
25. Kurzweil, P., Dietlmeier, O.K.: Elektrochemische Speicher. Springer Vieweg, Wiesbaden (2015)

26. Mitschke, M., Wallentowitz, H.: Dynamik der Kraftfahrzeuge. Springer Vieweg, Wiesbaden (2014)
27. Schütz, T.: Hucho—Aerodynamik des Automobils. Springer Vieweg, Wiesbaden (2008)
28. Hoepke, E.: Nutzfahrzeugtechnik. Springer Vieweg, Wiesbaden (2016)
29. Diercks, J.: Servicesicherung und Kostensenkung durch Logistik. Gabler Verlag, Wiesbaden (1985)
30. StVZO § 19 Abs. 2, StVO § 22 Abs. 1 and § 23 Abs. 1 (German Law)

Electrification of Urban Three-Wheeler Taxis in Tanzania: Combining the User's Perspective and Technical Feasibility Challenges

Mirko Goletz⊙, **Daniel Ehebrecht**⊙, **Christian Wachter**⊙, **Deborah Tolk**⊙, **Barbara Lenz**⊙, **Meike Kühnel**⊙, **Frank Rinderknecht, and Benedikt Hanke**⊙

Abstract This study assesses the feasibility of electric three-wheelers as moto-taxis in Dar es Salaam, Tanzania from a socioeconomic and technical point of view. The analysis is based on three pillars: (i) the acceptance of users (the moto-taxi drivers) for adoption, (ii) the vehicle specifications incl. battery type and size, and (iii) the role of the charging infrastructure. Findings are based on data from empirical field-work; methods used are qualitative and quantitative data analysis and modelling. Main findings include that moto-taxi drivers, who we see as most important adopters, are open towards electric mobility. They request however that vehicles should have similar driving characteristics than their current fuel-vehicles. As the market is very price sensitive, keeping the vehicle cost is of high importance. A high potential to lower these costs is seen by offering opportunity charging spots around the city. If such an infrastructure is being implemented the combination with suitable, cost competitive vehicles makes the transformation of the vehicle market towards electrification possible.

Keywords E-mobility · Charging infrastructure · Opportunity charging · Informal transport · Moto-taxi

M. Goletz (✉) · D. Ehebrecht · B. Lenz
DLR-Institute of Transport Research, Rudower Chaussee 7, 12489 Berlin, Germany
e-mail: mirko.goletz@dlr.de

D. Tolk · M. Kühnel · B. Hanke
DLR-Institute of Networked Energy Systems, Carl-Von-Ossietzky-Str. 15,
26129 Oldenburg, Germany

C. Wachter · F. Rinderknecht
DLR-Institute of Vehicle Concepts, Pfaffenwaldring 38-40, 70569 Stuttgart, Germany

© The Author(s) 2021
A. Ewert et al. (eds.), *Small Electric Vehicles*,
https://doi.org/10.1007/978-3-030-65843-4_8

97

1   Introduction

Two- and three-wheeled motorcycle-taxis (moto-taxis) have a significant and increasing modal share in rural and urban areas of developing countries, especially in Sub-Saharan Africa [1–3]. In urban areas, the popularity of the moto-taxis is linked to their ability to fill supply gaps in transport systems that are a result from fast urban growth and sprawl, insufficient urban and transport planning, and the overall deterioration of public transport in recent decades [4–6]. However, besides improving the mobility of citizens and providing job opportunities, combustion fuelled two- and three-wheelers increasingly contribute to air pollution and other negative externalities such as noise, traffic congestion and safety issues [2, 7–9].

In Dar es Salaam, largest city of Tanzania, the number of two- and three-wheelers has steadily grown in the last decade, and today they have become commonly visible in the streetscapes. Back in 2014, over 50,000 three wheelers were registered in Tanzania, and for 2018 WHO reported a total of 1,282,503 two- and three-wheelers in the country, 59% of the nation's registered vehicles [8]. In Dar es Salaam, moto-taxis form a *de facto* public transport system. They connect the main trunk roads with residential areas as well as feeding the public transport system [9, 10], often in areas where no other publicly available transport modes exist. Drivers of moto-taxis are commonly organized in groups, who wait at designated moto-taxi stands for their clients. Usually, drivers are only active at one or a small number of stands. Cruising is uncommon and drivers usually return to their stand after each ride. In 2010, the city administration legalized the operation of moto-taxis, but at the same time, restricted the access to the city-centre due to safety reasons and increased traffic congestion. Another argument was the need to improve air-quality.[1] Today, the city-centre lacks a public transport system for shorter distances to complement the bus rapid transit (BRT) that is designed to feed in commuters from the peri-urban areas. Along the BRT routes, especially the three-wheelers connect passengers to the BRT system (Fig. 1).

Being locally emission free, the electrification of the three-wheeler taxi-fleet represents a possible solution to face the sustainability challenges that could ultimately also allow moto-taxis to access the city centre again. It could also be a sustainable transport solution for other parts of the city. This study will evaluate the feasibility of electric vehicles as replacement for combustion-engine-driven moto-taxis in Dar es Salaam. The analysis is based on three pillars: (i) the acceptance of users (the moto-taxi drivers) for adoption, also with respect to the market conditions and business models and an assessment of the technical feasibility, with (ii) respect to the vehicle specifications such as battery type and size, and (iii) the role of the charging infrastructure. Our research is based on empirical field-work, data analysis, and modelling.

---

[1]Group discussion with public officials from "Surface and Marine Transport Regulatory Authority, (SUMATRA) during a stakeholder workshop at the University of Dar es Salaam in March 2018.

Fig. 1 Moto-taxi in Dar es Salaam. Photo taken during field trip Feb. 2019 © DLR/Benedikt Hanke

In result, we find a high level of acceptance of electric mobility amongst moto-taxi drivers, who currently own the vast majority of three-wheelers in Dar es Salaam. These findings however only hold if the electric vehicles fulfil the following criteria: driving characteristics (top-speed, passenger capacity) that are similar to the combustion fuelled three-wheelers, a sufficient range and cost competitiveness. To date, no such vehicles exist on the international three-wheeler market, and if, they would be too costly, mainly due to the battery cost that increase with capacity. This is an obstacle for a quick adoption. We therefore continue to analyse how battery sizing can be interlinked with charging infrastructure: If opportunity charging spots are installed at dedicated moto-taxi stands, the currently existing adoption barriers could be reduced. This could ultimately help to fasten adoption of electric three-wheelers in Dar es Salaam.

2 Methods and Data Collection

The study builds on a mixed methods approach that relies on empirical field work for data collection, modelling of the vehicle and power train as well as the battery storage and charging infrastructure.

2.1 Data Collection

Four methods for data collection were applied during two field stays in Dar es Salaam (November to December 2018, February 2019): (1) A paper-pencil survey among a total of 105 drivers at four different moto-taxi stands, located in the subwards of Mikocheni, Sinza and Mbezi was carried out. Questionnaire items included socioeconomic profiles of drivers, vehicle data, information on vehicle ownership, income and operation costs, as well as service characteristics such as the estimated number of trips per working day and regular operating hours. (2) Group discussions with representatives of two out of the four moto-taxi associations were conducted. The exchange served to assess the drivers' acceptance of E-mobility and to unveil potential benefits as well as obstacles linked to an implementation of electric three-wheelers in the city. (3) Explorative, open interviews with key stakeholders from public authorities, research institutions and private companies were conducted to assess regulatory frameworks and technical feasibility of implementing E-mobility in Dar es Salaam and to inquire about battery recycling capacities and the local vehicle market. (4) A GPS tracking campaign, over a period of one full week in each case, was carried out among three-wheeler taxi operators at the four different moto-taxi stands in order to gather vehicle driving profiles. The tracking was conducted with 1 s resolution in time using the DLR Moving Lab.[2] The drivers were asked to take notes for clarification of measurement artefacts and for additional information (passengers carried per trip, refuelling). During the measurement campaign 65 individual vehicle profiles, each representing one week of driving activity, with over 18 million individual data points covering 33,637 km were collected (see Fig. 2).

2.2 Modelling of Vehicle and Power Train

The design of the power train for an electric three-wheeler followed a multi-step approach and included the following steps:

1. Internet-based market research of available electric three-wheelers (carried out in January 2019) on a global scale, to understand what vehicles are available and asses if they are suitable to be used for the use-case as moto-taxi in Dar es Salaam.
2. A workshop with moto-taxi drivers (February 2019, c.f. 2.1) with the aim to identify needs, expectations and barriers about using electric three-wheelers.
3. Development of a power train concept reflecting the inputs from step 2.
4. Set up of a simulation model for three-wheelers with the power train concept developed in step 3 that simulates the energy demand.

---

[2]https://movinglab.dlr.de/en/.

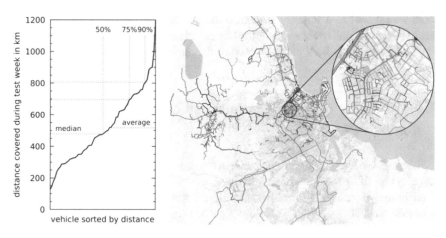

Fig. 2 Kilometres covered per vehicle during the test sorted by distance covered (left). Map of Dar es Salaam's city centre and it's outskirts with tracks of the GPS measurement campaign (right, figure uses map data from "© OpenStreetMap contributors, CC-BY-SA, www.openstreetmap.org/copyright")

Regarding the overall development of the vehicle, special attention was given to the need to keep the vehicle cost low (c.f. 3.1). The development steps therefore focused on the power train, which jointly with the battery is the main cost driver. In this step, two battery technologies were considered: lead acid and lithium ion batteries. Lead acid batteries are widely available in Tanzania as 12 V starter batteries. Therefore, choosing an intermediate circuit level of 48 V is reasonable. For lithium ion batteries, an intermediate circuit level of 48 V or 400 V can be considered, with 400 V being widely-used in electric vehicles, giving some advantage in the overall efficiency of the power train. Given that the authors rate safety a high issue, especially in developing countries, a 48 V intermediate circuit level is favoured.

The selection of the inverter technology depends on the intermediate circuit level: a MOSFET (metal-oxide-semiconductor field-effect transistor) inverter in case of 48 V, an IGBT (insulated-gate bipolar transistor) inverter in case of 400 V. Motor technologies can be rated under various aspects. Based on the findings of step 2 (c.f. above) low cost, longevity and easy maintenance were rated highest, while efficiency and power density were given lower priority. Consequently, we consider an asynchronous motor to be the best option, alongside to a gear box with a fixed ratio that allows for a smaller and cheaper overall system [11].

To calculate the vehicles energy demand, we set up a simulation model which uses efficiency maps or fixed efficiency values for the chosen power train components. Additionally, a drive cycle is needed like the Worldwide harmonized Light Duty Test Cycle (WLTC). However, after riding three-wheeler moto-taxis in Dar es Salaam during a field trip ourselves, we agreed that the real usage of those vehicles does not come close to the velocity profiles of standard drive cycles. Hence using them would lead to a falsified simulated energy demand. We therefore developed our own representative drive cycle for Dar es Salaam, using the GPS data from our empirical work (c.f. 2.1), partly following the approach of Eghtessad [12].

2.3   Modelling of Battery Storage and Charging Point Interdependency

In this step, we analyse the interdependence of a minimum required vehicle battery capacity, to fulfil the individual driving needs at minimal vehicle cost, and the availability of charging infrastructure at the dedicated moto-taxi stands. Our approach aims at describing a solution for a one-to-one replacement of combustion engine based vehicles (with current driving profiles) with battery electric vehicles. Based on the driver's interviews (c.f. 2.1), cost neutral solutions will have the highest acceptance for adoption. Based on the driving behaviours extracted from the GPS tracks of the tracking campaign, relevant charging locations at the moto-taxi stands were selected. Based on the availability of this charging network, we calculated the minimal required battery capacity for the individual vehicle.

We identified main moto-taxi stands of the drivers using a histogram methodology, creating a heat map of all measured vehicle locations. The identified stands matched with the findings from the driver interviews. Six public spots were identified as suitable public or semi-public charging locations to conduct so called opportunity charging. Opportunity charging in this study is defined as charging while waiting for customers during normal business hours at the point of business (here, specifically at the location of the dedicated moto-taxi stands mostly frequented).

The number of vehicles during a specific week of GPS tracking (c.f. 2.1) and the maximum number of vehicles at any station at the same time during that week is shown in Table 1. The coincidence factor is between 67 and 93%. If all vehicles need access to a charging point whenever they are at the station, up to one charging point per vehicle would be required. In the recorded data, most of the time, at least one vehicle is available at the main station of the driver group, with very few exceptions. Highest availability of vehicles is recorded around noon. Vehicle availability is higher during the day than during the night.

Table 1 Vehicles under test (VUT) and coincidence factor of moto-taxis at the stations

Week	# VUT	Max.	# VUT at station (%)
1	10	8	80
2	14	13	93
3	21	14	67
4	20	16	80

For the charging of the three-wheelers, we assumed that every parking event longer than 3 h is used to charge the vehicle, independent of the location. Introducing a technology change would incentivize drivers to select long term parking spots that would allow them to recharge the vehicles during longer breaks, where the vehicle is out of business. If a parking event takes place within 300 m of an opportunity charging station at the moto-taxi stands for more than 17 min, opportunity charging was conducted in the model. 12 min of the time at the stand are deducted for arriving, connecting and unplugging the vehicle. Taking these model assumptions into account, the shortest charging event lasts at least 5 min. Two charging scenarios with typical household currents of 7 A (1.6 kW) and 16 A (3.7 kW) and two SOC-target scenarios (80 and 100%) were considered.

During charging, an energy loss of 5% from grid to battery was taken into account. The technical battery capacity was assumed to be utilizable from 20 to 90%. The corresponding battery weight was included in the energy demand model of the vehicle (c.f. 2.2). The state of charge (SOC) was defined based on the effectively available (useable) battery capacity as 70% of the technical battery capacity. We applied a simplified battery model to account for the increased charging time demand above 80% SOC and conservative assumptions on the maximum charging rate where taken (see Eq. (1)).

$$C = \begin{cases} 0.5 & \forall SOC < 80\% \\ 0.2 & \forall SOC \geq 80\% \wedge SOC < 90\% \\ 0.1 & \forall SOC \geq 90\% \end{cases} \tag{1}$$

We conducted ensemble simulations considering different charging currents and vehicle loads to identify the range of energy storage demand in the vehicle battery with focus on the identification of the effectiveness of the availability of opportunity charging stations at the central moto-taxi stands and the robustness of the solution.

Further assumptions and thoughts: The economic and social framework in Dar es Salaam does not yet allow for higher-priced solutions like DC fast charging, battery swapping or inductive charging, as these solutions increase the upfront investment demand per operated vehicle. These technologies were not considered in this study. Technologically pragmatic and cheap to install charging opportunities at the central gathering points of the drivers could create job opportunities and reduce the battery storage demand for the vehicles, while significantly cutting the vehicle cost.

3 Results

3.1 Economic Feasibility and Drivers' Acceptance of E-Mobility

The socioeconomic survey we carried out among 105 three-wheeler taxi drivers was designed to assess economic feasibility of implementing electric three-wheeler taxis in the local context. In this regard, the results of the survey provide information on current service characteristics, vehicles and maintenance, vehicle ownership, costs and revenues, and socioeconomic data (Table 2).

The results show that the moto-taxi operators have long-time experience in the transport sector, working on average for 5.8 years as a taxi driver. Working times are long with drivers spending 6.4 days per week and 13.4 h per day on the job. The reason for long working hours is relatively high operating costs and the need to carry out as many passenger trips per day as possible in order to gain sufficient daily revenues that also allow supporting family dependents (avg. of 5 persons). A majority of 61% of interviewed drivers carries out more than 15 passenger trips per working day, while 15% attain over 25 passenger trips. Operating costs result not only from vehicle maintenance (avg. of 54,889 TZS/month, ∼24 US$) and fuel consumption (avg. of 11,822 TZS/working day, ∼5 US$), but mainly from the fact that a majority of drivers work on vehicle rent contracts (58%) or pay for a hire-purchase contract (6%). On average they spent 18,265 TZS (∼8 US$) each day on vehicle rent, independent from their working days, hours and revenues. As the gross income per working day amounts to 42,990 TZS (∼19 US$), it can be derived that most drivers gain relatively small net incomes.

Moreover, survey results indicate that drivers face social insecurities resulting from the threat of road accidents and necessary vehicle repairs, such as major engine failures (16.5% of vehicles affected). These incidents not only involve social and financial costs but can also lead to loss of income. Another aspect in regard of job insecurity is the seemingly high fluctuation in vehicle access: 63% of interviewed drivers gained access to their current vehicle only in 2018 or 2017, while most of them have been working in the sector for much longer than that. It is unclear if this high vehicle turnover results from frequently occurring terminations of rent contracts by vehicle owners or from unreliable or worn-out vehicles that need replacement: in this regard, results show that by the end of 2018 only a small share had a vehicle age of more than five years. In any case, buying or renting a new vehicle involves monetary and transaction costs that can put an extra burden on the drivers. The results consequently show that the moto-taxi sector is highly cost-sensitive—a fact that needs strong consideration when implementing E-mobility by replacing or retrofitting the existing vehicle fleet.

The aspect of cost sensitivity was also reflected in the group discussions with two moto-taxi stand associations: the drivers agreed that an E-mobility solution for three-wheeler taxis must not come with additional costs compared to current combustion fueled vehicles (which cost from 2800 to 3500 US$ according to local

Table 2 Summary of survey results among moto-taxi drivers ($n = 105$)

Socioeconomic data							
On average drivers are 34.6 years old				A driver's income supports 5 dependents			
Drivers have been working as moto-taxi operators for 5.8 years (median = 5)							
Working days and hours							
Drivers work 6.4 days/week				Drivers work 13.4 hours/day			
Number of passenger trips per day							
1–5	*6–10*	*11–15*	*16–20*	*21–25*	*26–30*	*>30*	
4%	17%	18%	29%	17%	7%	8%	
Year of vehicle production							
Unknown	*2012 or earlier*	*2013*	*2014*	*2015*	*2016*	*2017*	*2018*
19%	13%	11%	13%	11%	11%	11%	11%
Year of vehicle access							
2013 or earlier	*2014*		*2015*	*2016*		*2017*	*2018*
6%	8%		15%	8%		14%	49%
Vehicle ownership							
36% owner drivers			58% renters			6% hire-purchasers	
Costs and revenue							
Drivers spend 18,265 TZS/day on vehicle rent (median = 20,000)				Drivers spend 11,822 TZS/working day on fuel (median = 12,000)			
Drivers spend 54,889 TZS/month on **maintenance (median = 50,000)**				Drivers earn a gross income of 42,990 TZS/working day (median = 45,000)			
Repairs and accidents							
16.5% of vehicles had a major engine failure				21.8% of vehicles needed major repairs			
11% of vehicles have been involved in one or more road accidents							
Opportunities for night charging							
99% of night parking areas have electricity access							
66% of vehicles parked on protected private parking area							

stakeholders). Slightly higher vehicle costs would be acceptable, given that maintenance and energy costs turn out to be lower over the vehicle lifetime. Further preconditions are efficient vehicle charging times and easy access to charging stations or other charging opportunities. Additionally, sufficient travel speed (up to 90 km/h), a motor with higher power (current three wheelers with combustion engine offer 5.5–7.6 kW) and the ability to cover sufficient distances are important requirements for the drivers.

As part of the group discussions, the drivers also produced a list of potential benefits but also challenges they associate with an implementation of electric three-wheeler taxis in the local context: In addition to technical improvements and higher comfort of the service vehicles (e.g. less vehicle vibration, reduced noise, reduced need for spare parts), the drivers see potentials of higher incomes due to the symbolic value of modern technology attached to electric vehicles and its likely attractiveness to their customers. Stated disadvantages encompass low vehicle range that could limit opportunities to carry out passenger trips into peri-urban areas, lack of knowledge regarding electric vehicle maintenance, unclear availability of spare

Table 3 Results of the market analysis of three-wheeled vehicles (as of January 2019)

	Seats			Top speed in km/h			Battery tech.			Voltage level			
	3	4	5	25	45	<60	Pb	Li	n/a	48 V	60 V	72 V	n/a
Number of vehicles	4	2	7	8	2	3	7	3	3	9	1	2	1

parts in the country, higher vehicle purchasing costs, and a potential danger of increased accidents due to the low noise emission of electric vehicles. Another major concern was that the electric system could turn out to be unreliable during the rainy season. These insights are highly valuable as they indicate important criteria for drivers' acceptability of E-mobility and thus provide inputs for identifying sustainable technical solutions.

Looking at the electric three wheelers being currently available for purchase first thing to mention is their rapidly growing number in the recent years with manufacturers being mainly located in China and India. Since technical data is often hardly available, only a total of 13 vehicles could be investigated in this study (see Table 3).

Most of the vehicles are designed to carry five people (driver + 4 passengers) and offer a top speed of 25 km/h. They use mostly lead acid batteries and an intermediate circuit voltage level of 48 V. Battery capacity is often around 5 kWh, which results in a stated range of up to 100 km. This consequently relative low energy consumption is related to the low top speed of the vehicles. Motor power is mainly below 5 kW, mostly even below 2 kW with one exception at 7 kW.

Taking 6.4 working days per week into account, we found that 50% of the vehicles travel less than 75 km per working day and 75% of all vehicles travel less than 109 km. On average 81 km are covered during a working day (see Fig. 2, left). The available vehicles seem to cover a wide range of driving demands, though at an insufficient top speed.

Vehicles being offered in the European Union can cost from around 3360 US$[3] up to 16,800 US$, while vehicles offered in India are in a price range of 1904 US$ up to 4760 US$. However, since for most of the vehicles prices were not available and the market is assumed to change quickly, these prices can only give a rough indication.

3.2 Drive Cycle

The generated drive cycle uses the GPS tracking data as input, Fig. 3 shows the result. The length has been set to 1000 s to make it easily comparable to the WLTC.

---

[3]2019 yearly average exchange rates were used. 1.12 US$/€ and 2307.06 TZS/US$ [13, 17].

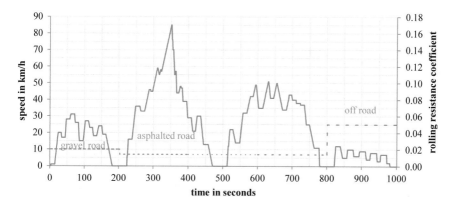

Fig. 3 Generated Dar es Salaam-drive cycle

It has been purposely separated into four sections which represent typical driving situations. To reflect the different road conditions, the rolling resistance coefficient is varied depending on the intended driving scenario of a section.

The first section represents a trip from a residential area to a main road, for example to feed the BRT system. The roads in residential areas are mainly small and gravel roads, which results in a maximum speed of 31 km/h. The second section is a trip on an asphalted main road from the suburbs to the city centre. Here, higher speeds up to 86 km/h can be reached. Despite having a very small portion in the measured trips, this maximum speed has been added on drivers' demand that wish for a fast vehicle and was the highest measured speed in the GPS data. The third section is a trip inside the city centre with a limited maximum speed of 52 km/h on asphalted roads. The last section is a trip through a residential area with a very bumpy dirt road. To model the higher energy demand for that kind of road, the rolling resistance coefficient has been significantly increased.

3.3 Energy Demand

Figure 4 shows the results for a simulated vehicle with full load (300 kg, equals approx. one driver plus three passengers) or half load (150 kg, one driver plus one passenger) performing the generated drive cycle (see Fig. 3). The simulation has been conducted for a vehicle using lead acid or lithium ion batteries which leads to a different overall weight of the vehicle. Battery size was varied in steps of 2 kWh in a range from 4 to 16 kWh.[4]

---

[4]This is not the technical but the usable battery capacity. To calculate battery weight, the energy density has been multiplied with 1.5 (lithium ion) or 1.1 (lead acid) to cover the additional weight of housing, cooling etc.

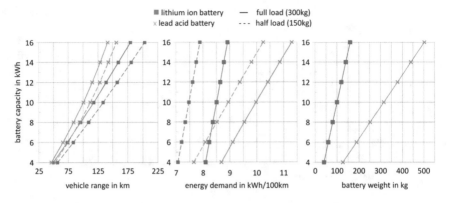

Fig. 4 Simulation results: range, energy demand, and battery weight for the generated drive cycle

Energy demand and vehicle range only differ slightly with increasing battery weight for the two variants which can be explained by the vehicles recuperation capability while braking. Though, the difference in battery weight widens strongly with increasing capacity.[5]

3.4 Battery and Opportunity Charging Point Demand

The battery demand of the vehicles to fulfil the mobility needs of the tested individual vehicles with and without opportunity charging is shown in Fig. 5 (left). Only lithium ion battery systems are shown, as we excluded lead acid batteries from the charging point analysis. We only assessed lithium ion batteries at this stage, due to the excess weight and negative environmental effects of lead acid batteries [14] and the expectation for a further price deduction of lithium ion batteries in the next years in.

The battery capacity demand distributions are fully separable for the cases with and without opportunity charging. To reduce the battery capacity demand of electric moto-taxis, the availability of opportunity charging stations at the moto-taxi stands is a viable measure. To serve 75% of the tested individual cases, the maximum calculated effective vehicle battery storage capacity is reduced from 22 to 12 kWh by introducing opportunity charging. Taking 2019 battery pack prices into account [15], a 12 kWh battery pack contributes at least 2016–4032 US$ to the vehicle price, while a 22 kWh battery contributes 3696–7392 US$—exceeding price expectancy ranges of the drivers and operators by far.

---

[5]For the lead acid variant, a stronger chassis would be needed, at least above a certain battery weight. This effect has not been included in the model and would widen the gap even more.

According to a distributor of three-wheelers (source: interview, c.f. Sect. 2.1), life times of three-wheelers in Dar es Salaam range between 3.5 and 6 years, driving 30,000 km per year and 100–200 km per day. If we assume a life time distance of 180,000 km, a battery capacity sufficient for 100 km requires 1800 full battery cycles, while 50 km require 3600 full battery cycles. Currently, lithium ion batteries reach 1500–2000 cycles, with 3000 cycles expected in 2030 [15]. The vehicle life of a battery electric three-wheeler with 50 km travel distance could therefore reach 150,000 km in 2030. Current studies suggest that a halving of today's prices until 2030 is likely to occur, with a vehicle life time of 150,000 km and prices between 84 US$/kWh [15] and 94 US$/kWh [16]. In this case, the price of a battery pack for a three-wheeler with 12 kWh technical battery demand would be between 1008 and 1128 $US while a 22 kWh battery would contribute 1848–2068 US$ to the vehicle price.

The charging point utilization at the defined stations per VUT is depicted in Fig. 5 (right) and maximum numbers are given in Table 4. The number of vehicles waiting at a station surpasses the number of vehicles actively charging most of the time. A full service availability scenario requires 0.43–0.80 charging points per vehicle, but only 0.29–0.70 charging points are utilized more than 1% (1 h 41 min) during the test week. As the batteries are charging faster from 0 to 80% SOC, the charging infrastructure requirements can be reduced by only charging the vehicle batteries to 80% SOC at the opportunity charging stations. This reduces the number of required charging points to 0.43–0.7 in a full service availability scenario, at the cost of a slightly higher battery demand (compare Fig. 5, left). The reduction of charging points and its effect on waiting time and moto-taxi service quality needs to be further studied in a field test, as the available data based on the behaviour of combustion engine driven vehicles is not suited to answer this question.

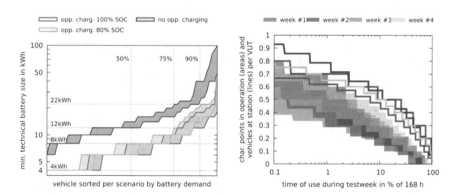

Fig. 5 Battery demand for battery electric vehicles with lithium ion batteries of the different scenarios with and without opportunity charging (left). Utilization of charging points as well as number of vehicles at stations normalized to the vehicles under test in a given test week (right)

Table 4 Vehicles under test (VUT) and charging points required for minimal battery size

Week	# VUT	# Charging points 100% SOC		# Charging points 80% SOC		# Charging points > 1% utilization	
1	10	7–8	70–80%	7	70%	4–7	40–70%
2	14	8–10	57–71%	8–9	57–64%	4–7	29–50%
3	21	9–14	43–67%	9–11	43–52%	6–11	29–52%
4	20	11–16	55–80%	11–13	55–65%	8–12	40–60%

4 Conclusions

The results of our study show that the electrification of three-wheeled moto-taxis in Dar es Salaam is possible, but subject to several constrains and barriers. Drivers show a high willingness to adopt electric vehicles; however, they require them to (a) at least deliver the same income than fossil-fuelled vehicles and (b) have similar driving characteristics. Besides, they expect electric three-wheelers to have a positive effect on the driving comfort for themselves and for their passengers.

From a socio-technological point of view, our analysis shows that currently available electric three-wheelers cannot be used as a one-to-one replacement. This is mainly due to the limited capability of the drive train, that limits the top speed and range that is requested by the drivers. Also, the current prices are not competitive.

Regarding technical specifications, safety issues lead us to recommend 48 V intermediate circuit voltage that is less likely to cause harm in case of maintenance and accidents. Lead acid batteries would be able to serve a capacity of up to 8 kWh, but due to the negative long-term environmental implications as a result from improper handling and disposal, lead-acid cannot be recommended. Above 8 kWh, due to their higher energy density, lithium ion batteries should be favoured.

Availability of infrastructure for opportunity charging reduces the demand for vehicle battery capacity. Charging infrastructure can therefore support the early adoption of electric moto-taxi by reducing overall vehicle cost. Charging points with household typical currents are sufficient to achieve this goal, with 0.43–0.7 charging points per vehicle being required.

Economic factors play an important role for the adoption of electric mobility, especially in the very cost sensitive Tanzanian vehicle market. Current lithium ion battery prices are too high for a wide adoption, but a price decrease is expected in the next 10 years. We therefore conclude that the combination of suitable, cost competitive vehicles and available opportunity charging infrastructure can make a transformation of the vehicle market towards electrification possible. In the future, policy makers should carefully reflect that changes to the structure of the informal moto-taxi market could require their interventions: In a market where driver organizations prevail, it can be assumed that these organizations will set up charging infrastructure, because this can help them reduce their cost and maintain the control to the access to the market. However, if technologies such as e-hailing (Uber, Bolt, SafeBoda) continue to take over market shares of the moto-taxi market [18], this may lead to a reduced amount of drivers who organize themselves in organizations and at designated stations. A publicly available charging infrastructure would then become more relevant for drivers who seek to buy electric vehicles, which policy makers should reflect when they want to speed up the adoption process.

The results we present here were generated for the city of Dar es Salaam. However, similar conditions, in which moto-taxis fill gaps of urban transport, exist in many cities of Sub-Sahara-Africa. Especially under similar driving patterns, our results are transferable to other places. We would expect that this is the case in cities where drivers are organized in organizations with designated stations and where morning and afternoon peak demand exists, allowing for opportunity charging with similar frequencies. Furthermore, the distances vehicles drive should be similar, which we expect to be the case when similar conditions related to land use and transport infrastructure are in place.

References

1. Diaz Olvera, L., Plat, D., Pochet, P., Sahabana, M.: Motorized two-wheelers in sub-Saharan African cities: public and private use. In: Presented at the 12th world conference on transport research, Lisbon, Portugal, 11–15 July 2010
2. Ehebrecht, D., Heinrichs, D., Lenz, B.: Motorcycle-taxis in sub-Saharan Africa: current knowledge, implications for the debate on 'informal' transport and research needs. J. Transp. Geogr. **69**, 242–256 (2018)
3. Starkey, P.: The benefits and challenges of increasing motorcycle use for rural access. In: Presented at the international conference on transport and road research, Mombasa, Kenya, 16–18 March 2016
4. Evans, J., O'Brien, J., Ch Ng, B.: Towards a geography of informal transport: mobility, infra-structure and urban sustainability from the back of a motorbike. Trans. Inst. Br. Geogr. **43**(4), 674–688 (2018). https://doi.org/10.1111/tran.12239
5. Klopp, J.: Towards a political economy of transportation policy and practice in Nairobi. Urban Forum **23**(1), 1–21 (2012). https://doi.org/10.1007/s12132-011-9116-y
6. Kumar: Understanding the emerging role of motorcycles in African cities. A political economy perspective. Sub-Saharan Africa transport policy program (SSATP), Discussion Paper 13. World Bank, Washington, DC, p. 9 (2011)
7. Adiang, C.M., Monkam, D., Njeugna, E., Gokhale, S.: Projecting impacts of two-wheelers on urban air quality of Douala, Cameroon. Transp. Res. D **52**, 49–63 (2017)
8. Global Status Report on Road Safety: World Health Organization—management of noncommunicable diseases, disability, violence and injury prevention (NVI), p. 258. Geneva, Switzerland (2018). ISBN 978-92-4-156568-4
9. Joseph, L., Neven, A., Martens, K., Kweka, O., Wets, G., Janssens, D.: Measuring individuals' travel behaviour by use of a GPS-based smartphone application in Dar es Salaam, Tanzania. J. Transp. Geogr. (2019). https://doi.org/10.1016/j.jtrangeo.2019.102477

10. Goletz, M., Ehebrecht, D.: How can GPS/GNSS tracking data be used to improve our understanding of informal transport? A discussion based on a feasibility study from Dar es Salaam. J. Transp. Geogr. (2018). Available online https://doi.org/10.1016/j.jtrangeo.2018.08.015 (in press)
11. Reupold, P.: Lösungsraumanalyse für Hauptantriebsstränge in batterieelektrischen Straßenfahrzeugen. Dissertation, Technische Universität München, München (2014)
12. Eghtessad, M.: Optimale Antriebsstrangkonfigurationen für Elektrofahrzeuge. Dissertation, Technische Universität Braunschweig, Braunschweig (2014). ISBN 978-3-8440-2782-2
13. Exchange USD—EURO: 1.12 \$/€, annual average 2019. https://www.statista.com/statistics/412794/euro-to-u-s-dollar-annual-average-exchange-rate/. Accessed 24 Apr 2020
14. United Nations Environmental Programme (UNEP): Lead acid batteries. https://www.unenvironment.org/explore-topics/chemicals-waste/what-we-do/emerging-issues/lead-acid-batteries. Accessed 30 Apr 2020
15. Ierides, M., del Valle, R., Fernandez, D., Bax, L., Jacques, P., Stassin, F., Meeus, M.: Advanced materials for clean and sustainable energy and mobility—EMIRI key R&I priorities. Energy Materials Industrial Research Initiative (EMIRI) Technology Roadmap, p. 292 (2019)
16. Battery requirements for future automotive applications. European Council for Automotive R&D (eucar), p. 4 (2019). https://www.eucar.be/battery-requirements-for-future-automotive-applications/. Accessed 29 Apr 2020
17. Exchange TZS—USD: 2307.06 TZS/US\$, annual average from 365 daily values. https://www.exchangerates.org.uk/USD-TZS-spot-exchange-rates-history-2019.html. Accessed 24 Apr 2020
18. Wadud, Z.: The effects of e-ride hailing on motorcycle ownership in an emerging-country megacity. Transport. Res. A: Pol. **137**, 301–312 (2020). ISSN 0965-8564. https://doi.org/10.1016/j.tra.2020.05.002

Impact Studies and Effects of SEV Deployment

Small Electric Vehicles (SEV)—Impacts of an Increasing SEV Fleet on the Electric Load and Grid

Tobias Gorges, Claudia Weißmann, and Sebastian Bothor

Abstract Heading towards climate neutrality, the electrification of the transport sector has significant impact on the electric grid infrastructure. Among other vehicles, the increasing number of new technologies, mobility offers, and services has an impact on the grid infrastructure. The purpose of this case study therefore is to examine and highlight the small electric vehicle (SEV) impact on the electric load and grid. A data-based analysis model with high charging demand in an energy network is developed that includes renewable energy production and a charging process of a whole SEV fleet during the daily electricity demand peak for the city of Stuttgart (Germany). Key figures are gathered and analysed from official statistics and open data sources. The resulting load increase due to the SEV development is determined and the impact on the electric grid in comparison to battery electric vehicles (BEV) is assessed for two district types. The case study shows that if SEVs replace BEVs, the effects on the grid peak load are considered significant. However, the implementation of a load management system may have an even higher influence on peak load reduction. Finally, recommendations for the future national and international development of SEV fleets are summarized.

Keywords Grid peak load · SEV · BEV · Electric grid · Grid stability

T. Gorges (✉)
MHP Management- und IT-Beratung, Berlin, Germany
e-mail: tobias.gorges@mhp.com

C. Weißmann
MHP Management- und IT-Beratung, Frankfurt am Main, Germany

S. Bothor
MHP Management- und IT-Beratung, Ludwigsburg, Germany

© The Author(s) 2021
A. Ewert et al. (eds.), *Small Electric Vehicles*,
https://doi.org/10.1007/978-3-030-65843-4_9

1 Introduction

Heading towards climate neutrality, which the German Government is aiming to achieve by 2050, the electrification of the transport sector has a significant impact on the electric grid infrastructure. Besides the increasing number of existing electric vehicles, new technologies, mobility services as well as small electric vehicles (SEV) need to be considered to assess the total impact on the grid infrastructure.

In the last few years, the fleet of SEVs increased by total number and type of vehicles. SEVs are emerging particularly in urban areas and are being more and more frequently rolled out as part of mobility-as-a-service approaches [1]. The growing SEV trend is predicted to continue up until 2030. Hence, this chapter will focus specifically on SEVs and their impact on the grid as well as the additional electricity demand and other factors of increasing grid stability uncertainty.

To examine and highlight the SEV impact on the grid infrastructure, the load capacity of a SEV fleet and its impact on the electric grid in 2020 and 2030 is calculated based on certain assumption in this study. Therefore, the charging process of the whole SEV fleet is modelled during the daily electricity peak demand. Vehicle replacement effects are taken into account and scenarios considering the fleet of battery electric vehicles (BEV) are modelled in this analysis.

To assess the SEV impact on the grid infrastructure, a data-based analysis model is set up and key figures like peak loads are gathered and analysed from official statistics and open data sources. As a study area the city of Stuttgart has been chosen, located in southern Germany and surrounded by a highly industrialized urbanized region with high electricity demand.

The determined charging capacity is compared with the load curve of a transmission system operator and the resulting load increase due to the SEV development is then calculated. Further, three scenarios are modelled by the modification of variables, hence the impacts on the grid infrastructure can be derived and recommendations for the future national and international development of SEV fleets can be drawn.

To conclude, this chapter derives answers to the following key questions: How much additional load capacity from SEVs considering the assumed development in the upcoming years can be expected by 2030? What are the impact differences between a purely BEV fleet and a mixed vehicle fleet of BEVs replaced by SEVs? Finally, how large is the difference in peak electric load between an SEV replacement in a BEV fleet versus load management?

2 Status Quo SEV and Grip Impact

Recent studies currently consider both mobility concepts and vehicle concepts when assessing ways to advance the transport system as a whole. Both concepts are driven by the current trends in mobility, such as: electric mobility, shared economy, digitalization, and autonomous driving [2–5].

However, the introduction of new concepts often is accompanied with challenges concerning the integration into the existing system such as the infrastructure. Due to this reason, an ever increasing amount of impact assessment research on the integration of new concepts in the mobility sector and their design is continuously published. The impact as well as the design of the proposed concepts is predominantly modelled. For instance, Alazzawi et al. determine the optimal fleet size of a shared autonomous vehicle concept for Milan [6].

The integration of electric mobility into the grid is one major research field. The central challenge of this field is: stabilization of the grid with consideration to the ranging power demand and production as well as additional spatial and time-related difference [7–10].

For instance, Morais et al. consider different scenarios of EV's impact in the power demand curve in a smart grid with vehicle-to-grid capabilities [11]. Bastida-Molina et al. assess charging strategies for light electric vehicles to exploit the existing electric generation system as best as possible [12]. Research on grid impact has predominantly been based on conventional-sized vehicle concepts and their battery characteristics. However, the impact and integration potential of different electrified vehicle concepts to ensure grid stability and especially peak load demand should have also been assessed.

Foresight, electrified autonomous vehicles and their already predicted impact on infrastructure, processes and culture [13, 14] would trigger a change of travel behaviour and travel choices [15, 16] to have a potential impact on SEVs. SEV, as new vehicle concept and pointed out in the preface and in the chapter "Small electric vehicles—benefits and drawbacks" of this book differ from BEVs, for example, in terms of their vehicle weight and thus battery and load capacity. Therefore, they have a certain potential for improving existing challenges of current transport systems. Hence, this chapter focuses on a prediction of the impact of an SEV fleet on the grid.

3 Method

The current available SEV concepts and their characteristics, such as seating capacity, maximum velocity, required driving licence, weather protection, predominant use (passenger or goods transport), volume of goods, determine the parameter set for the assessment of the potential for trip usage. This methodology is an adaptation of the approach used in the SEV study by Brost et al. [17] and data

Table 1 Overview of SEV use and development

Trip purpose	Trip distance (km)	Number of SEVs by 2020	Number of SEVs by 2030
Business	20.1	30	665
Shopping	9.1	58	1264
Leisure	13.0	38	823

retrieved from a mobility study in Germany (Mobilität in Deutschland (MiD), 2017) [18]. The identified trips with SEV potential are clustered by trip purpose. Mobility profiles throughout the day are developed serving as basis for the energy calculations. Besides the SEV fleet, the BEV fleet and its size in 2030 are calculated for the study location, Stuttgart in Germany. The results are integrated into a scenario analysis in chapter The UK Approach to Greater Market Acceptance of Powered Light Vehicles (PLVs).

Trip distances that are derived from the MiD are assumed to be equivalent in the future and differ between the three categorized trip purposes. It is assumed that an SEV can be assigned to one of these three predominant trip purpose categories, the total number of SEVs on each trip purpose is then determined. Numbers of SEVs in Stuttgart are predicted to rise up to a stock of 2800 by 2030 which equals more than 35,500 driven passenger-kilometres per day, as listed by Table 1. As only limited data on the SEV market and its future development is available, the SEV fleet size for 2030 is calculated using the same ramp-up time of BEVs.

In a second step, the corresponding electricity consumption and charging time is calculated in order to create a representative charging load profile for the scenario analysis (see Table 2).

The average electricity consumption of an SEV is 7.5 kWh/100 km and it is typically charged with a standard shockproof plug at 2.4 kW [17]. In comparison with this charging standard, an average BEV has a much higher consumption rate of about 20 kWh/100 km due to higher mass and speed [19]. BEVs can be charged via charging stations with higher loads.

In a third step, the calculated load profiles are applied to electricity grid branches for the assessment of the SEV impact. Different SEV usage scenarios and thus an increase of charging load are considered. As SEVs have a different impact on the grid than BEVs, the peak demand from SEVs and therefore necessary grid improvements and other measures are dependent on SEV usage.

Table 2 Overview of SEV electricity consumption and charging time

Trip purpose	Trip consumption (kWh)	Equivalent charging time (min)
Business	1.51	38
Shopping	0.68	17
Leisure	0.98	24

4 Case Study Stuttgart

Located in southwestern Germany in a valley basin, the city of Stuttgart contains a dense road network with a high amount of cars (301,793 vehicle registrations to 620,445 residents) [20, 21]. The shift towards electric mobility is ongoing with 3365 electric vehicles (battery electric and plug-in hybrid vehicles) on the road in 2019. Concerning the electric grid, about 90,000 power connections currently exist. About 97% of the power connections are realized using underground electric cables. This implies high costs in case of increasing the city's energy transport capacity due to necessary ground works.

To assess the impact of an SEV fleet on the grid, three different scenarios are created. Scenario 1, as the baseline scenario considers trips using BEVs only not including any SEV transportation. Scenario 2 is constructed around the assumption of trips using a fleet of SEVs and BEVs. However, here, the SEV trips do not replace the BEV trips but increase the total vehicle number on the road (worst case). In contrast to that, Scenario 3 describes a best-case trip scenario with an SEV fleet that is used for current car trips and can therefore replace some of the BEVs that would have otherwise been required on the roads.

The assessment focus for the impact analysis of SEVs on electricity grids in these scenarios is the change in peak demand in the city area as a whole and also in specific branches of the grid. In addition, the impact of the controllability of charging processes will be assessed.

Table 3 shows the parameters applied within the scenario analysis. The absolute number of BEVs describes the number of electric cars in Stuttgart including plug-in hybrids. In 2020, 3,365 BEVs exist in Stuttgart which corresponds to a local vehicle share of about 1% [20]. It is predicted that the BEV share will increase up to 24% or 73,514 by 2030 [22, 23] considering a total vehicle increase of about 1.2% due to urbanization [24]. The BEV calculation is described in detail in Weissmann and Gorges [23].

The scenario analysis focuses on the overnight charging process at a private charging point. According to the MiD study the average vehicle typically arrives at home at 08:00 PM and leaves at 07:00 AM, which determines the plug-in timespan. It is assumed that SEVs are charged with a standard shockproof plug at 2.4 kW while BEVs are charged via a 11 kW charging station (charging stations with higher delivery rates are neglected as they are not assumed to be economically

Table 3 Scenario parameters (without substitution)

Vehicle type	2020		2030	
	BEV	SEV	BEV	SEV
Absolute number	3365	126	73,514	2800
Maximum charging load (kW)	37,000	300	808,700	6600

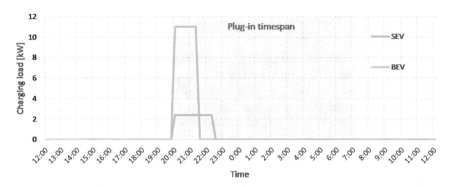

Fig. 1 Exemplary SEV and BEV charging profile (Charging after four "Business" trips, see Table 2)

efficient for the long overnight timespan) which can be controlled with a charging management system via a CPO-backend. Regarding controllability, this implies that the charging process of BEVs is controllable while the SEV charging is completely dependent on user behaviour.

For estimating the maximum possible charging load, a simultaneity factor of 1 has to be assumed for electric vehicles without charging management [25]. The overall effect of a charging management system on parallel charging cannot be quantified yet. Figure 1 illustrates an exemplary charging profile of a BEV and an SEV.

Based on these charging profiles and the scenario parameters of Table 3, maximum charging power loads are calculated. The results are shown in Table 4. SEVs have a total charging load of about 7 MW in 2030 (Scenario 2). However, assuming that SEVs will actually replace BEVs (Scenario 3) a total load reduction of approximately 24 MW could be achieved in 2030.

With the 11 kW charging technology for BEV charging, the increase of the maximum load up until 2030 is considerably high. Especially, if it is assumed that charging procedures will be started right at the beginning of the plug-in timespan (as in Fig. 2), the charging peak would correspond to the peak of the standard load profile of German households (H0 profile, see [26]). Considering an estimated peak demand of about 800 MW[1] of private households in the case study scope today, the total resulting peak demand could become twice as high by 2030 [27].

In order to derive the impact of additional charging loads on the city peak demand, two typical low voltage network branches are analysed. One branch supplies early adopters and therefore is located in an outlying district of the city with a small number of households per residential building in general. The other is assumed to be located in a central area of the city with residential buildings and a higher number of households per building. In the latter case, the rate of EV

---

[1]390,000 electric meters × 10 kW peak load per meter × 0.2 simultaneity factor.

Table 4 Maximum charging load—Scenario perspective

Scenario	Scenario 1	Scenario 2	Scenario 3
	BEVs only	BEVs plus SEVs	SEVs replace BEVs
Maximum grid charging load 2020 (kW)	37,000	37,300	35,900
Maximum grid charging load 2030 (kW)	808,700	815,300	785,000

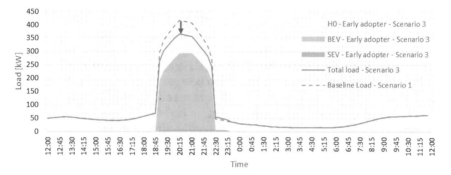

Fig. 2 Total load reduction in Scenario 3 compared to Scenario 1 (Early adopter, outlying district)—H0 BDEW household load profile according to [26]; Distribution of charging timespan according to [28]

adoption is considered to be lower. Another main difference of these two branches is the number of cars per inhabitant. There are 424 cars per 1,000 inhabitants in the early adopter's branch, which is significantly lower than the 307 cars in the inner city, with a city wide average of 365 cars per 1,000 inhabitants [24].

Table 5 illustrates the possible increase of the peak load for both branch cases in each scenario. In Scenario 1, the baseline BEVs-only scenario, the increase of the peak demand is generally high. In addition, the socio-demographic difference between the two cases is shown as the peak demand increases about 50% in the average adopter street and almost three times as high in the early adopter street.

Table 5 Peak demand increase in hypothetical branches*—Scenario perspective

Scenario for 2030	Scenario 1	Scenario 2	Scenario 3
	BEVs only (%)	BEVs plus SEVs (%)	SEVs replace BEVs (%)
Increase of peak demand—average adopters street, inner city	48	52	30
Increase of peak demand—early adopters street, outlying district	190	197	167

*Based on the worst-case assumption of simultaneity charging factor of 1

In contrast to this large difference, the results in Scenario 2 are almost similar to those of Scenario 1. This is a result of the assumption that SEVs are charged with only 22% of the charging power used for BEVs. Therefore, the increase of the peak demand in a typical branch can be almost totally explained by the charging of BEVs. Hence, an additional use of SEVs as assumed in Scenario 2 does not have a major additional negative impact on the grid.

The focus of Scenario 3 is if SEVs could have a positive effect on the grid instead, by replacing trips made by BEVs. A maximum substitution rate of 4% of the BEVs is assumed (see Table 3). Table 5 shows that due to substitution, especially in the average adopter street in the inner city a significantly lower peak demand with only 30% increase is calculated. On the other hand, the early adopter street in the outlying district a very high range of additional peak demand (167%) might still occur due to electric vehicle extension.

Figure 2 depicts the possible load reduction in Scenario 3 for this early adopter branch in comparison to the baseline Scenario 1. Here, a peak load reduction of about 50 kW can be achieved by SEV substitution (see red arrow). These effects of substitution are considered significant and will have not only an overall, but also various local neutralizing impacts. Lower investments in grid developments might therefore be sufficient in some streets and grid branches and less total restructuring necessary as well as less civil engineering interference with daily city life. As a result, advantages will occur within grid branches with higher SEV substitution levels.

On the other hand, there will still be a high number of branches requiring grid development due to the strong BEV uptake. Even if SEVs replace 20% of BEV trips, a very high level of additional load is still likely to be served through the physical electricity grid of the city branches. As a conclusion, SEVs can only partly neutralize intense grid peak demand situations in future.

However, one should keep in mind that the peak load values are derived using a worst case assumption with a simultaneity factor equal to 1 for charging. With future charging management systems, the controllable BEV loads could be distributed over longer timespans (until 7:00 AM in the morning) which would result in a possible BEV peak load reduction of approximately 70% (the blue BEV loads within the 3.5 h charging time-span could be equally distributed over the plug-in timespan of 11 h, see Fig. 2) as well as in a total load reduction. Looking at Fig. 2, one can conclude, that the grid advantages due to charging management could be higher than the impact due to SEV trip replacement, which reduces peak load by 15%.

5   Conclusion

The aim of this chapter has been to assess the impact of SEVs on the electric grid, assuming that the electric grid will be stressed by electric mobility in general in the future. In chapters Courses of Action for Improving the Safety of the Powered Cycle and Velomobiles and Urban Mobility: Opportunities and Challenges, the status quo of SEVs is described. Feasible SEV trip purposes have been identified to derive the potential for SEVs to replace BEVs.

A scenario analysis has been conducted for the city of Stuttgart to calculate the SEV grid impact in chapter The UK Approach to Greater Market Acceptance of Powered Light Vehicles (PLVs), focusing two city districts with different socio-demographic conditions. Within the scenario scope, a BEV-only extension (as a baseline) has been compared to a BEV increase with additional SEVs and a BEV increase utilizing SEVs to replace a certain percentage of trips.

The analysis shows that the impact of electric vehicles on the total grid load in general is very high, implying the requirement for essential measures to secure grid stability in the long run. The negative effect of additional SEV load is however negligible (Scenario 2) as the occurring electric vehicle peak loads in 2030 are mainly driven by BEV charging. This is a result of the much higher absolute number of BEVs compared to SEVs as well as much lower load peaks of singular SEVs (2.4 kW) compared to singular BEVs (11 kW). Moreover, SEVs may have a positive effect on grid stability when used to replace BEV trips (Scenario 3). Especially, in outlying city districts with early adopters to electric mobility and likewise an accelerated BEV and SEV increase, this effect is significant. Nevertheless, the analysis also has pointed out that charging management as another grid stabilization measure may have a higher impact (70% BEV peak load reduction) on total peak load reduction than SEV substitution effects as the assumed share of SEVs in 2030 is still rather low.

There are certainly some limitations of this chapter due to a lack of SEV data in comparison to existing BEV data. Nevertheless, the authors conducted this study as accurately as possible with the available data. With an increasing data quantity and quality on current and future SEV fleets, the assumptions of the chapter, such as the defined vehicle characteristics and fleet size, as well as their impact on the electric grid could be verified and assessed in more detail respectively. For example, there could occur a much higher SEV extension in reality, due to the fact that SEV extension is not linked to the development of proper charging infrastructure like BEV extension.

Another aspect that should be considered in future research is the low controllability of the SEV charging processes via shockproof wall sockets. Based on the assumption that SEV will be mainly charged at low power (3.7–7.2 kW) at wall sockets, charging controllability is only possible if there is an available control signal from the grid operator or aggregator. As charging controllability is a prerequisite for the best utilization of BEV local or wider grid services, just as congestion management or ancillary services are, the SEV contribution to this type of grid stability measure is currently little to none. A possible improvement for SEVs could be an adaptation of BEV backend technology to control the charging process by grid operator signals. With a significant SEV fleet, likewise another positive grid effect could be created.

Finally, as the number and popularity of SEVs increases, more information on the usage of SEVs should become available (such as specific trip characteristics and changing mobility behaviour) and could be considered in future research. Aligned with the changing mobility sector in general, this, however, will be a continuous research topic.

References

1. Giesecke, R., Surakka, T., Hakonen, M.: Conceptualising mobility as a service. In: 2016 11th International conference on ecological vehicles and renewable energies, EVER 2016, January 2018 (2016). https://doi.org/10.1109/EVER.2016.7476443
2. Burns, L.D., Jordan, W.C., Scarborough, B.A.: Program on sustainable mobility the Earth Institute, Columbia University, pp. 1–43 (2013)
3. Fagnant, D.J., Kockelman, K.: Preparing a nation for autonomous vehicles: opportunities, barriers and policy recommendations for capitalizing on self-driven vehicles. Transp. Res. **77** (Part A), 167–181 (2015)
4. Fagnant, D., Kockelman, K.M.: The travel and environmental implications of shared autonomous vehicles, using agent-based model scenarios. Transp. Res. Part C **40**, 1–13 (2014). https://doi.org/10.1016/j.trc.2013.12.001
5. Jaekel, M., Bronnert, K.: Die digitale Evolution moderner Großstädte (2013). https://doi.org/10.1007/978-3-658-00171-1
6. Alazzawi, S., Hummel, M., Kordt, P., Sickenberger, T., Wieseotte, C., Wohak, O.: Simulating the impact of shared, autonomous vehicles on urban mobility—a case study of Milan. 2, 94–76 (2018). https://doi.org/10.29007/2n4h
7. BDEW: Elektromobilität braucht Netzinfrastruktur. Netzanschluss und-integration von Elektromobilität. https://www.bdew.de/media/documents/Stn_20170615_Netztintegration-Elektromobilitaet.pdf (2017)
8. Bundesnetzagentur: Bericht zum Zustand und Ausbau der Verteilernetze 2018 (2018)
9. Friedl, G., Walcher, F., Stäglich, J., Fritz, T., Manteuffel, D.: Blackout—e-mobilität setzt Netzbetreiber unter Druck (2018)
10. Nobis, P., Fischhaber, S.: Belastung der Stromnetze durch Elektromobilität. Belastung Der Stromnetze Durch Elektromobilität, **3** (2015)
11. Morais, H., Sousa, T., Vale, Z., Faria, P.: Evaluation of the electric vehicle impact in the power demand curve in a smart grid environment. Energy Convers. Manag. **82**, 268–282 (2014). https://doi.org/10.1016/j.enconman.2014.03.032
12. Bastida-Molina, P., Hurtado-Pérez, E., Pérez-Navarro, Á., et al.: Light electric vehicle charging strategy for low impact on the grid. Environ. Sci. Pollut. Res. (2020)
13. Bauer, G.S., Greenblatt, J.B., Gerke, B.F.: Cost, energy, and environmental impact of automated electric taxi fleets in Manhattan. Environ. Sci. Technol. **52**(8), 4920–4928 (2018). https://doi.org/10.1021/acs.est.7b04732
14. Sprei, F.: Disrupting mobility. Energy Res. Soc. Sci. 37, 238–242 (2018). https://doi.org/10.1016/j.erss.2017.10.029
15. Fraunhofer IAO, Horváth Partners: The value of time—Nutzerbezogene Service-Potenziale durch autonomes Fahren, vol. 80, issue April (2016)
16. Soteropoulos, A., Berger, M., Ciari, F.: Impacts of automated vehicles on travel behaviour and land use: an international review of modelling studies. Transp. Rev. **39**(1), 29–49 (2019). https://doi.org/10.1080/01441647.2018.1523253
17. Brost, M., Ewert, A., Schmid, S., Stieler, S., Eisenmann, C., Klauenberg, J., Gruber, J.: Elektrische Klein- und Leichtfahrzeuge - Chancen und Potentiale für Baden-Württemberg on behalf of e-mobil BW (2019)
18. Nobis, C., Kuhnimhof, T.: Mobilität in Deutschland—MiD Ergebnisbericht (2018)
19. ADAC: Aktuelle Elektroautos im Test: So hoch ist der Stromverbrauch. https://www.adac.de/rund-ums-fahrzeug/tests/elektromobilitaet/stromverbrauch-elektroautos-adac-test/ (2020). Accessed 31.07.2020
20. Kraftfahrbundesamt: Bestand am 1. Januar 2019 nach Zulassungsbezirken und Gemeinden. Bestand an Kraftfahrzeugen Und Kraftfahrzeuganhängern Nach Zulassungs-bezirken. https://www.kba.de/DE/Statistik/Fahrzeuge/Bestand/ZulassungsbezirkeGemeinden/zulassungsbezirke_node.html (2019). Accessed 31.07.2020

21. Statistisches Amt Stuttgart: Fortgeschriebene Einwohnerzahl in Stuttgart 2019 nach Staatsangehörigkeit und Geschlecht. https://www.stuttgart.de/item/show/688662/1 (2019). Accessed 31.07.2020
22. Gerbert, P., Herhold, P., Burchardt, J., Schönberger, S., Rechenmacher, F., Kirchner, A., Kemmler, A., Wünsch, M.: Klimapfade für Deutschland. In: BDI—Bundesverband der Deutschen Industrie e. V., BCG—The Boston Consulting Group, Prognos. https://bdi.eu/publikation/news/klimapfade-fuer-deutschland/ (2018). Accessed 31.07.2020
23. Weissmann, C., Gorges, T.: Entwicklung eines Planungstools zum regionalen Ladeinfrastrukturausbau. In: Realisierung Utility 4.0, Band 2: Praxis der digitalen Energie-wirtschaft von den Grundlagen bis zur Verteilung im Smart Grid (1. Auflage). Springer Vieweg (2019)
24. Statistisches Amt Stuttgart: Statistikatlas Stuttgart. https://statistik.stuttgart.de/statistiken/statistikatlas/atlas/atlas.html?indikator=i0&select=00 (2019). Accessed 31.07.2020
25. DIN 18015-1: Elektrische Anlagen in Wohngebäuden—Teil 1: Planungsgrundlagen (2018)
26. Meier, H., Fünfgeld, C., Adam, T., Schieferdecker, B.: Repräsentative VDEW-Lastprofile. Forschungsbericht (1999)
27. Stuttgart Netze GmbH: Elektromobilität in Stuttgart nutzen. https://www.stuttgart-netze.de/netz-nutzen/themen/elektromobilitaet/ (2020). Accessed 31.07.2020
28. Eckstein, S., Buddeke, M., Merten, F.: Europäischer Lastgang 2050. Projektbericht zum Arbeitspaket 4: Regenerative Stromerzeugung und Speicherbedarf in 2050—Restore 2050 (2015)

Fields of Applications and Transport-Related Potentials of Small Electric Vehicles in Germany

Christine Eisenmann, Johannes Gruber, Mascha Brost, Amelie Ewert, Sylvia Stieler, and Katja Gicklhorn

Abstract The possible applications of small electric vehicles, i.e., electric cargo bikes and three- and four-wheeled L-class vehicles in transport, are discussed, and potential business models are presented. Moreover, transport-related potentials are analyzed. Therefore, we have utilized a multi-method approach: we conducted qualitative interviews with experts and professionals in the field of light and small electric vehicles and carried out quantitative analyses with the national household travel survey mobility in Germany 2017. Our results show that, theoretically, small electric vehicles could be used for 20–50% of private trips (depending on the model). On these trips, however, they would not only replace car trips, but also trips on public transport or by bicycle and on foot. In commercial transport, these vehicles are particularly suitable for service trips and some last-mile deliveries. If small electric vehicles were to replace a significant share of the transport volumes of motorized passenger and commercial transport, they could contribute to climate protection.

Keywords Small electric vehicles · Mobility in Germany MiD · Transport impacts · User potentials · LEV applications

C. Eisenmann (✉) · J. Gruber
German Aerospace Center (DLR), Institute of Transport Research, Rudower Chaussee 7, 12489 Berlin, Germany
e-mail: Christine.Eisenmann@dlr.de

M. Brost · A. Ewert
German Aerospace Center (DLR), Institute of Vehicle Concepts, Wankelstaße 5, 70563 Stuttgart, Germany

S. Stieler
IMU Institut, Hasenbergstraße 49, 70176 Stuttgart, Germany

K. Gicklhorn
e-Mobil BW GmbH, Leuschnerstraße 45, 70176 Stuttgart, Germany

© The Author(s) 2021
A. Ewert et al. (eds.), *Small Electric Vehicles*,
https://doi.org/10.1007/978-3-030-65843-4_10

1 Introduction

Climate change, congestion, air pollution, and increasing transport volumes are putting pressure on cities and municipalities worldwide to enhance sustainable transport and mobility. Politicians, decision makers, and transport researchers are aiming to reduce land use, satisfy people's desire for mobility and goods availability, and improve air quality and the quality of life in cities. Small and light electric vehicles (SEVs) are seen as a feasible factor in meeting these challenges [1–3]. The global market is developing dynamically in regard to SEV sales, particularly in Asian countries like China and India [4]. In Europe, however, sales are very low, and SEVs are not widely available [5]. Moreover, at present, the framework conditions in Germany might hinder a broader market adoption of SEVs [6]. Applying supporting measures could foster a more widespread use. Therefore, it is important to analyze the possible potential of SEVs in respect to their impact on transport. Against this background, our contribution addresses the following three topics:

1. Transport impacts: What is the share of transport demand in passenger transport and commercial transport that could be substituted by SEVs? On which trips could SEVs be used?
2. User potentials: For which user groups are SEVs suitable in terms of socio-demographic characteristics?
3. Applications and business models: What are the possible applications of SEVs in passenger transport and commercial transport? Which business models might be conceivable and suitable for SEVs?

SEV Definition The presented analyses focus on three-wheeled (L2e and L5e) and four-wheeled (L6e and L7e) SEVs according to EU regulation No. 168/2013 as well as electric cargo bikes. Cargo bikes can be two-, three- or four-wheeled with and without insurance tags.

2 Materials and Methods

We utilized a multi-method approach to answer our research questions: we conducted qualitative expert interviews amongst SEV experts and professionals and carried out quantitative analyses with national household travel survey (NHTS) data. We have furthermore compared and contrasted our results with scientific literature. For the present publication, the NHTS data analysis is focused on Germany. The expert interviews, in contrast, refer to the German Federal State Baden-Württemberg appropriate to the primary study.

2.1 Qualitative Expert Interviews

SEVs are rarely used in Germany. Hence, statistics on the, e.g., employment effects and economic impacts of SEVs in Germany do not exist yet. In order to broaden our findings from literature and travel survey analyses, we have conducted eleven expert interviews. We developed an interview guideline, which we have used as a basis for each interview. For a comprehensive view, the interviewed experts cover a wide range of expertise: from entrepreneurs of small, medium-sized, and global corporations to scientists and representatives of the state government of Baden-Württemberg. This included expertise in the production and sale of SEVs as well as the assessment of the legal framework in the study. The interviews were recorded and analyzed for congruence. Our conclusions are based on according statements. The content-related statements from the experts helped to evaluate the quantitative findings from data analysis and to formulate recommendations for action. In accordance with the sponsor (e-mobil BW GmbH) of the study, the interviews serve for a first exploration of the research field.

2.2 Quantitative Data Analyses with the NHTS Mobility in Germany

The NHTS Mobility in Germany (MiD) The MiD is a nationwide, comprehensive survey on travel behavior and transport demand of the German residential population [7]. The current survey was conducted in 2017; former surveys date back to 2002 and 2008.

The MiD is a one-day survey, i.e., each participant reported their mobility on a given day, i.e., the survey day. The field phase of the MiD 2017 took place between May 2016 and September 2017. Therefore, a mixed-mode approach was applied: the participants took part in the survey either by paper and pencil questionnaire or by web questionnaire. A total of around 316,000 individuals from 156,000 households participated and reported 961,000 trips on their respective survey days.

The MiD 2017 survey offers various levels of data analysis. The central component is the trip data set, which contains all trips made by the survey participants on their survey days. This data set contains, for example, trip-specific information on departure and arrival time, purpose, modes of transport, distance travelled, and accompaniment by other persons.

If the survey participants make regular trips due to their profession (e.g., mail carrier, craftsmen, bus drivers, elderly care services), information on these trips was collected at a lower level of detail (e.g., only the total distance travelled of all business-related trips on the survey day).

It should be noted that no conclusions can be drawn from the one-day survey data of the MiD 2017 on the travel behavior of individuals and SEV user potential over longer periods of time. Consequently, no conclusions can be drawn from the MiD information about SEV user potentials of single individuals from a longitudinal perspective. However, the analyses allow to quantify the share of the reported trip chains for which SEVs are suitable according to a defined set of requirements and trip chain specifications. A trip chain is a sequence of trips that start at one place (often at home) and end there again. Thereby, a technical potential of using SEVs can be shown.

Estimation of the Maximum Feasible Transport and User Potentials of SEVs For our analysis, we have selected seven characteristic SEV models with heterogeneous vehicle characteristics (e.g., electric range, maximum speed). Those vehicle characteristics were used to determine the characteristics of trip chains on which SEVs can be used. If a trip chain meets all requirements, it is assumed that an SEV can theoretically be used on that trip chain. Tables 1 and 2 provide a summary of the considered SEV models.

With our approach, we are able to determine the maximum feasible potential of SEVs. However, we need to note that the maximum potential shown will never be fully realized. This is because even if there is a large supply of SEVs, not every person will use their SEVs for every trip that they could have taken an SEV on. Other aspects, such as personal preferences or willingness-to-buy, as well as economic feasibility, also impact individual mode choices. For example, many transport users with short commutes could cycle to work, but do not do so because of personal preferences. These aspects were not addressed in the MiD 2017 and can therefore not be included in our analysis.

Vehicles used for commercial transport (e.g., service trips of craftsmen or some types of deliveries) are often provided by the employer. Therefore, SEVs for private use (passenger transport) and SEVs for commercial transport are considered in separate analyses. The data basis for the analysis of private trips is the travel diary of the MiD. The data basis for the analysis in commercial transport is the less detailed dataset of "regular business-related trips" (rbW) of the MiD; therefore, rbW are used as a proxy for commercial transport within this publication.

The results for passenger transport and commercial transport are comparable as the estimations are based on the same survey and method. This was seen as more fruitful than to use the slightly outdated KiD 2010 survey (Motor Vehicle Traffic in Germany 2010), which has been frequently used for empirically-based researched in commercial transport.

In order to determine trip chains for which SEVs could be used, various aspects are examined in a differentiated procedure for each individual trip chain in the MiD. Only if all aspects are applicable, it is assumed that an SEV could be used on this trip chain. We have considered the following aspects:

- Is the electrical range of the SEV sufficient for the distance covered?
- If the person is accompanied by others: Does the SEV offer additional seats?
- If goods need to be transported: Does the SEV have facilities that allow the transport of such objects?

Table 1 Vehicle and usage characteristics of various SEV models, suited primarily for passenger transport

	EVT trike	Riese & Müller Packster 80 HS	Aixam eCity Pack	Micromobility systems Microlino
Range (km)[1]	70	63	75	140
Seats	2	1	2	2
Number of wheels	3	2	4	4
Goods volume (l)	Approx. 10	135	700	300
Maximum speed (km/h)	45	45	45	90
Driving license required	Yes	Yes	Yes	Yes
Weather protection	No	No	Yes	Yes

[1]Practical range estimated on the basis of information provided by manufacturers

Table 2 Vehicle and usage characteristics of various SEV models, suited primarily for commercial transport

	Riese & Müller Packster 80 HS	Kyburz DXP 4	Radkutsche	Alkè ATX 320E
Range (km)[1]	63	50	68	52
Seats	1	1	1	2
Number of wheels	2	3	3	4
Goods volume (l)	135	Variable	Variable	Variable
Maximum speed (km/h)	45	45	25	44
Driving license required	Yes	Yes	No	Yes
Weather protection	No	No	No	Yes

[1]Practical range estimated on the basis of information provided by manufacturers

– Is the permissible maximum speed of the SEV coherent with the road infrastructure that was used on the trip chain?(Estimation based on the average speed
   on the longest trip of the trip chain)
– If a driving license is required for the SEV: Does the survey participant have the
   appropriate driving license?
– Does the SEV have sufficient weather protection to enable its use even in
   unfavorable weather conditions?

A separate analysis is carried out for each of the seven SEV models in Tables 1
and 2. Since the data collection of regular business-related trips is less detailed, only
the first two questions can be examined for analysis in commercial transport. We
have also taken into account that trip chains should be at least 800 m long, as
shorter trip chains do not justify the time required of access and egress of SEVs.

3   Results

The multi-method approach with qualitative expert interviews, quantitative analyses
of national household travel survey (NHTS) data, and analysis of scientific literature provides findings on application of SEVs, technical potential of trip or,
respectively, transport share which is described in the following sections.
Thereafter, we present applications and business models for SEVs. The underlying
study of this book chapter evaluates a variety of additional SEV topics including
measures that could be implemented for fostering SEVs [8].

3.1   Feasible Transport Impacts and User Potentials of SEVs
        in Passenger Transport

Depending on the respective SEV model, a maximum of 17% to 49% of all private
trips and 6 to 30% of the distance covered by private trips can be substituted by
SEVs, see Fig. 1.[1] The electric range and maximum speed in particular limit the
substitution potential here.

SEVs with higher ranges and speeds are a feasible alternative, especially on
commuting trips and shopping trips: the Microlino could be used on 57% of all
commuting trips and on 59% of all shopping trips. Figure 2 shows for which share
of trips SEVs could be used, which have so far been covered on foot, by bicycle,
motorized private transport (MIV), or public transport. Thus, SEVs represent a
certain rivalry to environmentally friendly transport modes. Vehicle models with

---

[1]We would like to point out that individual trips in a trip chain could also have been carried out
with an SEV (e.g., when using SEV sharing services or when recharging the SEV during the trip
chains), but this was not analyzed here.

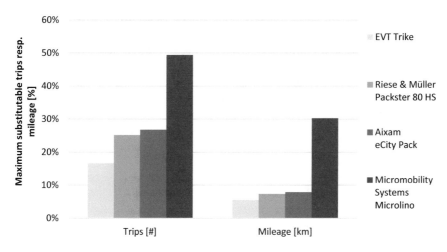

Fig. 1 Share of private trips and mileages on private trips that can be substituted by SEVs (maximum potential)

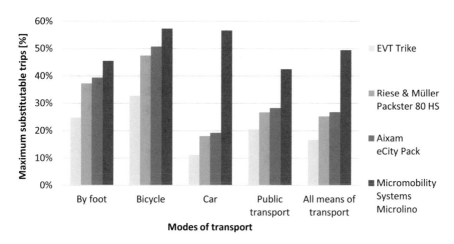

Fig. 2 Proportion of private trips that could be substituted by SEVs, differentiated by previous modes of transport (maximum potential)

lower ranges and maximum speeds are technically suitable for a higher share of trips covered by active modes than trips covered by car. For example, the EVT Trike could be used for a quarter of all trips on foot and a third of all trips by bicycle. In contrast, the EVT Trike could only replace 11% of car trips and one fifth of all trips by public transport. However, these numbers do not indicate that there is a high risk of replacing environmentally friendly modes of transport as the survey data does not contain information on user willingness to change transport modes. SEV models with larger electric ranges and maximum speeds, such as the Microlino on the other hand, are better suited to replace MIV trips instead: After all, 57% of MIV trips and 42% of public transport trips could also be made with the Microlino. Regarding active modes of transport, the Microlino is technically suitable for 46% of walking and 57% of cycling trips. A major reason why SEVs cannot be used on more walking or cycling trips is that some of the SEVs considered require a driving license. Not all persons who travelled on foot or by bicycle on the survey day have the appropriate driving license. Although trips with active modes are usually within the electric range of SEVs, no higher degree of substitutability is feasible because either a driving license is often required to use the SEV or trip chains on foot are shorter than 800 m.

Moreover, the question arises as to which groups of people could use SEVs in their everyday travel and how large the potential user group is. For this purpose, the MiD 2017 was used to analyze which share of the population on the survey day made trips for which an SEV could also be used, and how these population groups are characterized.

Between 16 and 38% of the participants in the MiD 2017 survey could have used an SEV on the survey day. These numbers again represent the technical usage maximum without taking personal preferences into account. It can be seen that SEV models with a higher electric range, and maximum speed have more potential to be used on private trips.

There is user potential for SEVs in all age groups from 18 years onward, with the user potential of the age groups from 30 years and older is higher than that of 18–29 year olds (see Fig. 3). The figure shows only persons aged 18 and over, since the MiD does not differentiate between different driving license classes and therefore does not include driving license classes with a minimum age of under 18 years.

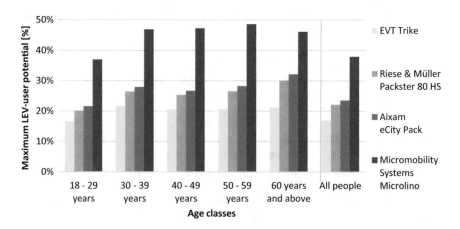

Fig. 3 Persons who could have used SEVs on the survey day, differentiated by different SEV models and age groups (maximum potential)

3.2 Feasible Transport Impacts and User Potentials of SEVs in Commercial Transport

SEVs can be used for about 34–56% of regular business-related trips, which are used as proxy for commercial transport (see Fig. 4). Due to the heterogeneity of commercial trips with a substantial amount of daily trip chains of 100 km or more; however, only about one tenth of the mileage could technically be replaced. The results for the KYBURZ DXP 4 are significantly lower in comparison with the other three SEV models. These findings represent an average; depending on the business sector, the percentages can be considerably higher, such as in postal delivery.

A further differentiation of the maximum substitutable trips was made with respect to the commercial trip purposes, such as business meetings, service trips, mobile nursery, or goods deliveries. This differentiation shows a relatively high degree of stratification for different dimensions of commercial transport (see Fig. 5). For the purpose of social service, healthcare, and nursery a maximum of 80% of the trips could be replaced by SEVs (in relation to the top-three SEV models that show a similar performance). For the purpose of customer service and business-related errands, a substitution level of 60% was found; similarly for visits, inspections, and meetings (substitution level some 50%). Note that, the two previous groups of trip purposes might contain private companies as well as local authorities or municipal companies. The lowest maximum substitution rate of about 40% was found for transport, pick-up, and delivery of goods, which roughly corresponds to the courier, express, and parcel logistics services (CEP) industry but might also include more general types of delivery services. Looking at these results, SEVs seem to be more promising in business sectors where the transport of goods is

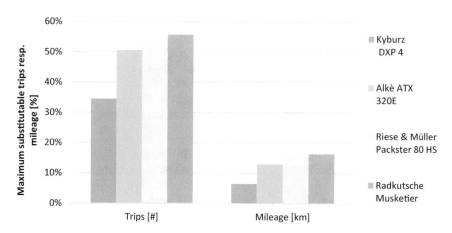

Fig. 4 Share of commercial transport trips and mileages on commercial trips that can be substituted by SEVs (maximum potential)

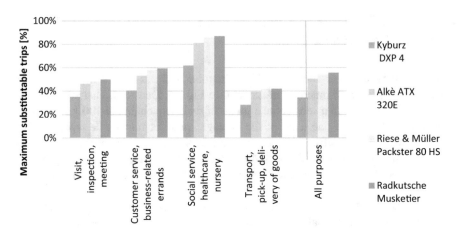

Fig. 5 Proportion of commercial transport trips that could be substituted by SEVs, differentiated by trip purposes (maximum potential)

not the core business but an additional requirement to fulfill the core business (e.g., any kind of mobile service).

Particularly in the case of freight transport, the limited data basis must be taken into account: If empirical data on actual payload requirements were available, the results would certainly lead to a further differentiation between the two differently-sized electric cargo bikes Musketeer (a heavy-load tricycle) and Packster (a "Long John" two-wheeler). Due to the vehicle design of the evaluated SEVs, only the SEV Alkè ATX is available for the transport of passengers, but due to the low general substitution potential, its use for this purpose is likely to be largely limited to locations such as airports or exhibition centers.

3.3 Feasible Applications and Business Models of SEVs

In passenger and commercial transport, there are various areas of application for SEVs, in line with the heterogeneous travel behavior of individuals and their private and professional activities. The interviewed experts see great opportunities in particular with regard to climate protection and the threat of driving bans for diesel and gasoline vehicles in city centers.

SEV Applications in Commercial Transport SEVs could be used commercially for CEP services, other delivery services, internal/on-site transports, and service trips.

Within CEP services, SEVs would rather be of implemented for courier deliveries, i.e., urgent B2B shipments delivered point-to-point by messengers. Technically, it is also conceivable to use them for the delivery of parcels to end-users (B2C). However, currently only a few of the available SEVs, such as the ALKE ATX, offer a sufficient carrying capacity. It is therefore less likely that SEVs will be used to a large degree in the parcel mass market, given the current regulation.

In contrast, SEVs could be used more frequently for other time-critical delivery services, especially for the delivery of prepared meals, as only a modest transport capacity is required here. Furthermore, SEVs could be used by retailers (or their contractors) for instant or time-window delivery options. However, due to low willingness-to-pay by customers, the feasibility of this business model might be a challenge.

SEVs are also suitable for internal or on-site commercial transports; this includes transport on larger company sites or residential facilities as well as transport between different sites of an organization or local authority. As these trips are often scheduled regularly with a homogeneous volume of goods, they can be adapted more easily to the limited ranges and loading capacities of SEVs.

Service trips are a heterogeneous group of trips within commercial transport in which the primary focus is not the transportation of goods, but the execution of services at the point of destination. These services require the transport of tools, spare parts, or other working materials. Some SEVs are well suited for service trips that only require moderate payloads and daily mileages, such as city cleaning and gardening, janitorial and facility services, technicians, craftsmen or nursing services.

SEV Applications in Passenger Transport In passenger transport, SEVs could be used for everyday mobility in the same way as other privately owned means of transport, such as the car. SEVs could, if the model characteristics allow it, be used on entire trip chains, e.g., from home to work, then to sport and back home in the evening. Particularly on shorter trip chains, SEVs are a comfortable transport solution independent of the present topography and often have weather protection. Many SEV models also offer the possibility of transporting goods, making them suitable for (smaller) shopping. Due to the lower land usage compared to the car, SEVs are also an interesting alternative means of transport in urban areas with limited parking space.

Another area of application is as mobility option for tourists. Especially in car-free tourism areas, e.g., nature parks, SEVs are an alternative to the bicycle or car for the mobility of tourists.

Moreover, SEVs could also be used in passenger transport on trip stages. This is feasible, for example, on the first or last mile of a public transport trip, i.e., on the stage between the public transport stop and the start or destination of the trip. In such an area of application, it would be conceivable that SEVs could be offered in sharing concepts. Similar to current bike-sharing concepts, parking facilities could be installed at public transport stops, so that a simple and quick transfer to and from public transport is guaranteed.

Table 3 Sharing services with SEVs in Europe

	Region	Name	Operators	Further information
	Metropolitan region, urban (Grenoble, FR)	Citélib by Ha:mo	Toyota, Metropolitan region, Municipality, Electricity supplier, Car sharing Company	Pilot phase (3 years) with 35 COMS, 35 i-Roads, 120 charging stations
	Urban (Paris, Clichy, FR)	Moov'in. Paris by Renault	Renault, ADA	Followed after ending of Autolib Car sharing, 20 Twizys and 100 Zoés
	Metropolitan region, urban (Ruhr, DE)	RUHRAUTO	Research Institute, Public Transport, Manufacturers, Municipalities	Research project with a diverse portfolio of vehicle models (incl. Twizy)
	Urban (various cities, I)	Share'ngo	Car sharing company	Currently operating in several Italian cities
	Urban (Biel, CH)	ENUU	Start-up, cooperation with municipality	Financing through app and advertising space on the vehicle
	Urban (Prague, CZ)	Re.volt		Fleet consists of 20 SEVs, electric motor bikes and e-scooters

Business Models for Mobility Services A potential business model in passenger transport is the use of SEVs in shared fleets. Potential users can test and get to know SEVs without obligation. Such concepts are already in place in some European cities, see Table 3. The design of those sharing concepts varies. In some cases, SEVs are offered exclusively (e.g., Citélib by Ha:mo, Re.volt), in other cases further vehicle categories are available in addition to SEVs (e.g., cars and commercial vehicles). Users can thus choose between various vehicle categories depending on the purpose of the journey and their needs. This enables flexible mobility. Manufacturers acting as sharing providers often want to increase the use and awareness of their own vehicle models among the population. Other providers establishing and operating sharing fleets are municipalities, energy suppliers, public transport facilities, car sharing providers, or start-ups.

Sharing fleets with SEVs are at present still a niche market. Success on the demand side depends on various factors, for example, the attractiveness of the service, usage costs, local availability of the service, and ease of use. The creation of significant demand is fundamental to the development of financially successful business models. Furthermore, when selecting SEV models for sharing, it is important to ensure that they are robust in use and resistant to external damage, so that the maintenance and servicing costs remain manageable. The resale and residual value of SEVs from sharing fleets in the secondary market determine the success of the business model. Moreover, one sharing fleet operator pointed out in our interview that the requirements for the robustness and reliability of the vehicles when used in sharing fleets are higher than for private use. This should be taken into account when designing SEVs. At present, no conclusions can be drawn regarding the profitability of this business model. Sharing services offered by car manufacturers with fleets of cars are currently aimed primarily at offering their own mobility services, strengthening the bond with their own customer base, and thus reducing competition from other mobility-service providers [9]. Furthermore, the strong growth in subscribers is giving the impression of a strong growth in demand for sharing services.

4 Conclusions and Outlook

Our results show that, technically, SEVs could be used for 20–50% of private trips (depending on the SEV model and only counting trips that are parts of trip chains, if SEVs are suitable for the entire trip chain). This technical potential includes car trips, as well as trips on public transport or by bicycle and on foot. The potential is restricted by the criteria used for the analysis such as availability of weather protection and average speed on the longest trip of the trip chain. Extended future analysis could also analyze single trips in order to account for options such as car sharing and recharging during trip chains, thereby deriving an even higher potential. Within commercial transport, SEVs have been found to be particularly suitable for service-oriented activities such as healthcare, craftsmen, or municipal services. Concerning deliveries, time-critical point-to-point (courier) shipments or food deliveries seem to be more feasible than standard parcel deliveries.

If SEVs were to replace a significant share of the transport volumes of motorized passenger and commercial transport, they could contribute to climate protection. Due to the lower weight and lower speeds of SEVs compared to cars, less energy is required for their operation.

The presented analysis of the MiD was accompanied by expert interviews that included questions regarding any possible measures to foster SEVs. Many experts stated that, if a wider application of SEVs with the replacement of specific modes is desired, boundary conditions need to be actively shaped accordingly. Since SEVs have already been on the market for more than 12 years without gaining a high market share, significant shifts in transport modes seem unlikely unless parameters, such as regulation, taxes, boundary conditions of use, vehicle technology, variety of models or prices, change fundamentally.

Future studies and real-world settings could evaluate how measures could foster SEVs as part of a sustainable transport system. According to the authors, measures should focus on areas of application, where public transport and active modes cannot offer attractive solutions, e.g., due to very low or disperse transport demand or to the physical constraints of users.

References

1. Honey, E., Lee, H., Suh, I.-S.: Future urban transportation technologies for sustainability with an emphasis on growing mega cities: a strategic proposal on introducing a new micro electric vehicle statement. WTR (3), 13 (2014). https://doi.org/10.7165/wtr2014.3.3.139
2. Santucci, M., Pieve, M., Pierini, M.: Electric L-category vehicles for smart urban mobility. Trans. Res. Part F Traffic Psychol Behav **14**, 3651–3660 (2016)
3. Bauer, W., Wagner, S., Edel, F., Stegmüller, S., Nagl, E.: Mikromobilität. Nutzerbedarfe und Marktpotenziale im Personenverkehr. Stuttgart, Germany: Fraunhofer-Institut für Arbeitswirtschaft und Organisation IAO (2017)
4. International Energy Agency: Global EV Outlook 2019: Scaling-Up the Transition to Electric Mobility. OECD (2019)
5. ACEM: Motorcycle, Moped and Quadricycle Registrations in the European Union—2010–2018. (2019). Accessed 08 Mar 2019. [Online] Available: https://www.acem.eu/market-data
6. Ewert, A., Brost, M.K., Schmid, S.A.: Framework conditions and potential measures for small electric vehicles on a municipal level. WEVJ **11**(1), 1 (2020). https://doi.org/10.3390/wevj11010001
7. Nobis, C., Kuhnimhof, T.: Mobilität in Deutschland—MiD Ergebnisbericht. Study from infas, DLR, IVT and infas 360 on behalf of the German Ministry of Transport and Digital Infrastructure.
8. Brost, M., Ewert, A., Schmid, S., Eisenmann, C., Gruber, J., Klauenberg, J., Stieler, S.: Elektrische Klein- und Leichtfahrzeuge. Chancen und Potenziale für Baden-Württemberg on Behalf of E-mobil BW. Stuttgart, Germany (2019)
9. Zeit Online: Carsharing: BMW und Daimler stecken Milliardenbetrag in Mobilitätsfirma. ZEIT ONLINE, 22 Feb 2019. https://www.zeit.de/mobilitaet/2019–02/carsharing-bmw-daimler-investition-mobilitaetsfirma-elektroautos. Accessed 30 Apr 2019

Vehicle Concepts and Technologies

KYBURZ Small Electric Vehicles: A Case Study in Successful Deployment

Erik Wilhelm, Wilfried Hahn, and Martin Kyburz

Abstract This paper is written from the perspective of a Swiss OEM which has been active in the small electric vehicle (SEV) market since 1991 and has put over 22,000 SEVs on the road around the world. KYBURZ Switzerland AG identified several important niche markets for SEVs and today sells vehicles to improve the mobility of senior citizens (e.g. KYBURZ Plus), to increase the efficiency of postal and logistics companies (e.g., KYBURZ DXP), and to imbue drivers with passion for electric vehicles (e.g., KYBURZ eRod). Most KYBURZ vehicles are currently homologated in the category L2e, L6e, or L7e. The company has also developed a Fleet Management product which gives its customers detailed insights into the performance of their electric as well as conventionally powered vehicles. Anonymized datasets from this Fleet Management system will be drawn upon in this paper to examine questions regarding their application, i.e., environmental and economic aspects. The unique feature which the authors from KYBURZ bring with this paper is that all their investigations are performed with real data gained from the field experience. The primary focus of this paper is on last-mile mobility services for postal organizations which help to increase efficiency and meet sustainability goals.

Keywords Electric postal vehicles · Large-scale deployment · Total cost

1 Introduction

The Swiss OEM KYBURZ offers a Fleet Management product which gives its customers detailed insights into the performance of their electric as well as conventionally powered vehicles. Anonymized datasets from this Fleet Management system will be drawn upon in this paper to examine the following questions:

E. Wilhelm (✉) · W. Hahn · M. Kyburz
Kyburz Switzerland AG, Solarweg 1, Freienstein-Teufen 8427, Switzerland
e-mail: erik.wilhelm@kyburz-switzerland.ch

© The Author(s) 2021
A. Ewert et al. (eds.), *Small Electric Vehicles*,
https://doi.org/10.1007/978-3-030-65843-4_11

1. How do small electric vehicles (SEVs) compare against incumbents across a variety of environmental performance indicators based on real-world operating data?
2. What advantages do SEVs present considering increasing urbanization and e-commerce, particularly in terms of urban delivery vehicles?
3. Which economic indicators are decisive for fleet customers when considering the switch to SEVs, with an emphasis on total cost of ownership?
4. Do differences exist between different geographic markets for SEVs which should be considered, particularly regarding the electrical grid supply mix?
5. What can be done at the end of life for SEVs which further increase their economic value and reduce the demand for critical raw materials?
6. How can telematics data from SEVs be effectively collected, anonymized, and analyzed to maximize their impact?

2 Methods

KYBURZ Switzerland has developed the Fleet Management infrastructure shown in Fig. 1 which enables vehicle operators and managers to collect, process, store, and operationalize a wide variety of light as well as heavy-duty vehicle data. In this section, we will address question six: *How can telematics data from SEVs be effectively collected, anonymized, and analyzed to maximize their impact?*

The KYBURZ Fleet Management server infrastructure is protected by state-of-the-art approaches to ensure that user privacy is not compromised by maintaining high levels of cyber security (TLS everywhere), in addition to complying with the General Data Protection Regulation (DSGVO). We use several

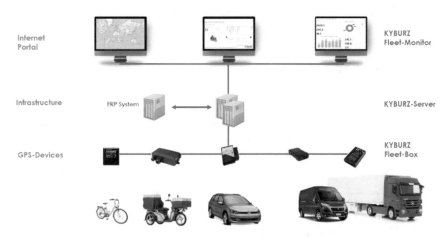

Fig. 1 KYBURZ Fleet management infrastructure for securely storing and processing data

Fig. 2 Distribution of daily driving distance for the week of the 10th to the 14th of June 2019

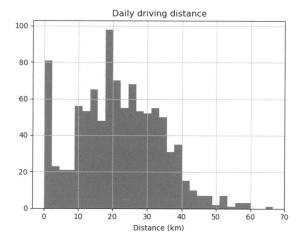

approaches to minimize the risk while collecting telematics data. For example, we do not link datasets collected from vehicles to datasets containing personal or corporate information which is stored on the ERP server illustrated in Fig. 1, and we used hashed identifies where possible.

The Fleet Management infrastructure collects data from over 2,500 DXP vehicles in daily postal operation. For the analysis presented in this chapter, a subset of N = 394 was selected over the week of the 10th to the 16th of June 2019. These vehicles travel a mean of 22 km/day during their postal rounds, distributed as shown in Fig. 2. The vehicles typically finish their tours with over 50% of battery capacity remaining.

3 Environmental Impact and Geography

In this section, we propose answers to the questions: *How do SEVs compare against incumbents across a variety of environmental performance indicators based on real-world operating data? Do differences exist between different geographic markets for SEVs which should be considered, particularly regarding the electrical grid supply mix?*

Starting in early 2010, the Swiss Post began to replace their fleet of gasoline two-stroke delivery scooters with KYBURZ Switzerland's SEVs both shown in Fig. 4 and described in Table 1. The present deployment of roughly 6,000 SEVs is shown in Fig. 3 and resulted in a reduction in fleet size of over 1,500 fewer last-mile delivery vehicles which brought important economic and logistical efficiency gains [2]. A service network (shown as purple dots in the map) consisting of roughly 60 partners maintains and manages the fleet. The vehicles are serviced every 5,000 km, and the deployment has resulted in a diminished absolute total cost of fleet ownership for the Swiss Post.

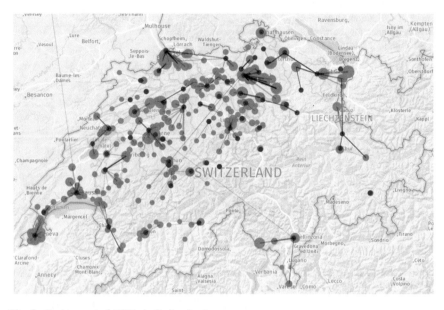

Fig. 3 Deployment of DXPs in Switzerland

Fig. 4 Piaggio Liberty 125 cc internal combustion engine (ICE) incumbent delivery vehicle (left) and the KYBURZ DXP electric delivery vehicle (right)

Table 1 Physical characteristics of the delivery vehicles

Parameters	Internal combustion	KYBURZ DXP SEV
Empty weight (kg)	95	210
Maximum load capacity (kg)	89	270

Using the input data and methodology from Sacchi [6] and Cox [3] as well as the input assumptions summarized in Table 2, we obtain the results in Table 3. Here, it can be clearly seen that the DXP SEVs offer substantial benefits over the incumbent

Table 2 Life cycle modeling inputs for the DXP SEV

	Best case	Mean case	Worst case
Lifetime (in km)	40,000	35,000	30,000
Lifetime (in years)	7	7	7
Mass driver (in kg)	70	70	70
Mass cargo (in kg)	125	198	270
Curb mass (in kg)	210	210	210
Of which, glider mass	138	138	138
Of which, electric motor mass	20	20	20
Of which, battery mass	52	52	52
Battery change during lifetime	No	Yes	Yes
Driving mass (in kg)	478	405	550
Motor peak power (in kW)	2.4	2.4	2.4
Electricity consumption (in Wh/km)	69.4	88.8	111.1
Battery capacity (in kWh)	4.5	4.8	5.2
Max discharge rate (in % of capacity)	80	80	80
Range (in km)	59	43	33
Swiss electricity CO_2 intensity (gCO_2-eq./kWh)	106	106	106

Table 3 Life cycle modeling results for the DXP SEV, compared to the incumbent ICE the DXP exhibits far superior performance

gCO_2-eq./km	Glider	Powertrain	Energy storage	Maintenance	Energy chain	Direct emissions	Road	Total
Best case EV	28	10	16	4	8	0	6	71
Mean case EV	32	11	36	4	10	0	6	100
Worst case EV	37	13	42	5	12	0	6	116
ICE	11	9	0	11	42	217	6	296

internal combustion engine (ICE) delivery vehicles, even when considering a conservative estimate of the electrical grid CO_2 intensity of 106 g CO_2/kWh. As expected, the major penalty incurred by the incumbent ICE is during the use-phase.

The results of analyzing the on-road energy use of the DXPs in various geographical markets under both summer and winter conditions is shown in Fig. 5, where it is clear that the electrical delivery vehicles offer over $8\times$ efficiency advantages versus the internal combustion engine incumbents, which as previously discussed in the LCA analysis drives the substantially lower total emissions. In postal delivery applications with a high loads and large number of starts and stops, the real-world ICE consumption is typically much higher than when tested to standardized cycles. In postal applications, the inherent efficiency of SEVs is a strong advantage.

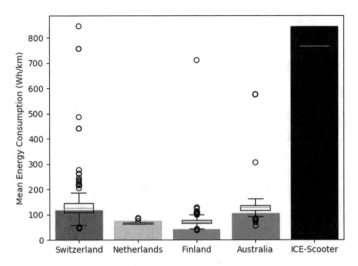

Fig. 5 Measured on-road energy consumption of the DXP in various markets compared with the ICE incumbent, showing a significant: energy-saving potential by switching to electric delivery vehicles

In a further analysis, we evaluated the CO_2 intensity of the electricity grid mix, and conclude that the DXP will either cut the CO_2 emissions by one third over its lifecycle, or in the best case reduce them by 50% depending on whether the primary energy comes from a natural-gas intensive grid or from the Swiss grid mix. Other recent studies have arrived at similar conclusions, for example, the work of Edwards [4].

4 Urban Space Implications

In this section, we answer the question: *What advantages do SEVs present considering increasing urbanization and e-commerce, particularly in terms of urban delivery vehicles?*

The shrinking volume of letter mail leaves a Post two choices: either reduce the mail network ("downward spiral") or maintain the network by adding items from other areas (e.g., small parcels). The mail network typically shows significant lower cost per item delivered compared to a parcel delivery van. Further, with the increase in e-commerce, the last-mile parcel delivery hits capacity limits. Going from door to door with a delivery van brings additional traffic congestion and emissions to city areas, whereas the mail network already uses small vehicles in cities. With SEV like the KYBURZ DXCargo, the mail network can easily be extended to deliver small parcels on a postman's letter route, keeping the cost per item delivered to a minimum. On the other hand, the parcel van can operate more efficiently focusing on bigger items with higher single value and less e-commerce packets.

Fig. 6 Compare the KYBURZ DXCargo with conventional panel delivery trucks

To examine the potential for SEVs such as the **KYBURZ** DXCargo to solve pressing urban logistics problems presented by conventional panel trucks (both shown in Fig. 6), we perform a simple line calculation to examine the space used by both delivery vehicles in this section.

The DXCargo has a maximum carrying capacity of 200 kg and occupies a footprint of 2.4 m^2. The conventional cargo van has a carrying capacity of 1,100 kg and occupies a footprint of 16.8 m^2. As such, to carry the same number of parcels by weight a postal service would require six DXCs with an aggregated footprint of 14.4 m^2. If six DXCs are deployed to deliver the same number of packages as one cargo van, cumulatively they use almost 100 m^2 less urban area over the course of driving a single route with an average of 50 stops relative to when a cargo van is used. Occupied surface per route is calculated by multiplying the number of stops times the footprint of the vehicle (fleet). Considering that 247,000 cargo vans were sold in 2016 [7], this sums to a traffic-reducing space saving of over 20 km^2 per day if all of the cargo vans sold in 2016 were replaced by SEVs like the DXCargo. Put in other terms, 2,869 soccer field sized areas would be saved in urban spaces through this swap. This is a conservative estimate, since it is likely more than 250,000 cargo vans are deployed for daily deliveries. A geospatial study of urban deliveries found that the distance walked to delivery packages was almost equivalent to the distance driven. Hence, we are comfortable making this head-to-head comparison [1].

Another important consideration is the time saved during delivery stops. A typical delivery route recorded using the KYBURZ Fleet Management system is shown in Fig. 7. Although it does not seem like much time, the eight seconds saved using three-wheeled SEVs like those produced by KYBURZ illustrated in Fig. 9 can sum to major cost savings [5]. The eight seconds of efficiency exist because the letters can be gathered while the brake is being applied, there is no need to extend or retract a side stand, forward, and reverse directions can be controlled simply and easily with a simple button click, and the DXP has improved agility (Fig. 8).

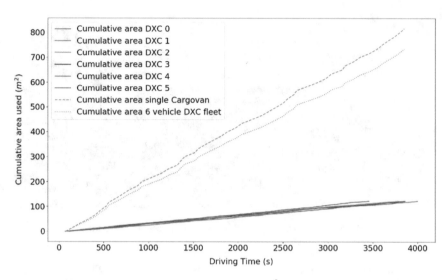

Fig. 7 Cargo van-based deliveries occupy almost 700 m^2 area per route versus DXC-based deliveries with one vehicle. A DXC fleet with six vehicles, capable of delivering the same amount of cargo as one big van, would save 100 m^2 per route

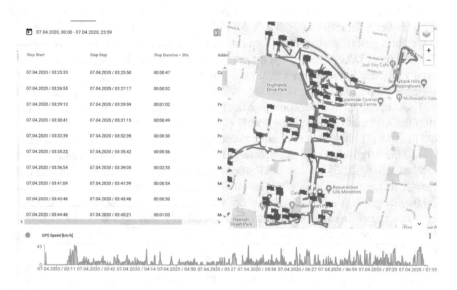

Fig. 8 A typical delivery route shows over 50 stops in postal service per day

Based on the daily summary of April 6th, 2020, one DXP fleet consisting of 1,124 vehicles made an average of 57 stops per day as can be seen in Fig. 10. On this basis, this fleet operator saves 4,528 working days per year for this relatively modest fleet of 2,000 vehicles.

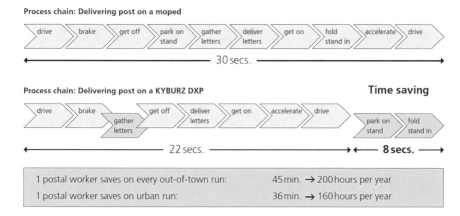

Process chain: Delivering post on a moped

| drive | brake | get off | park on stand | gather letters | deliver letters | get on | fold stand in | accelerate | drive |

← 30 secs. →

Process chain: Delivering post on a KYBURZ DXP **Time saving**

| drive | brake | gather letters | get off | deliver letters | get on | accelerate | drive | | park on stand | fold stand in |

← 22 secs. → ← 8 secs. →

| 1 postal worker saves on every out-of-town run: | 45 min. → 200 hours per year |
| 1 postal worker saves on urban run: | 36 min. → 160 hours per year |

Fig. 9 Average time savings recorded for real postal routes between services using conventional versus three-wheeled SEVs for delivery

Fig. 10 DXP vehicles stop an average of 57 times per route, resulting in substantial time and cost savings

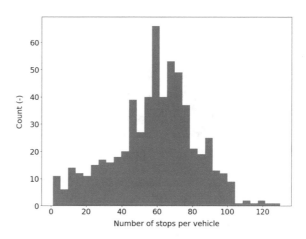

Count (-)

Number of stops per vehicle

5 Commercial Efficiency

Here, we examine the question: *Which economic indicators are decisive for fleet customers when considering the switch to SEVs, with an emphasis on total cost of ownership?*

Considering only purchase, maintenance, fuel, and disposal costs based on the assumptions illustrated in Table 4, we analyze the total cost of ownership for last-mile delivery fleets. We assume that both types of vehicles must travel the same km per year, although it is likely that the internal combustion engine (ICE) vehicles would be required to travel further due to their diminished load capacity. We also include a positive residual value of 700 € for the SEV due to SecondLife considerations described in Sect. 5.

Table 4 Assumptions for calculating the total cost of ownership comparison

	Electric delivery vehicle	ICE incumbent
Purchase price CHF	10,000	3,000
7 year replacements	1	2.5
Carrying capacity kg	270	89
Km/year	5,000	5,000
Fuel cost CHF/L-eq	0.65	1.49
Maintenance cost CHF/km	0.29	0.24
End of life disposal cost	−700	100
Total cost of ownership CHF/km	0.57	0.59

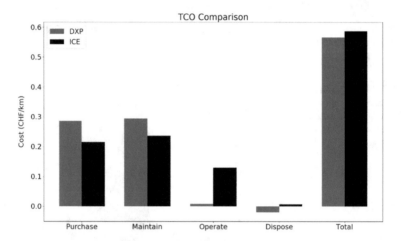

Fig. 11 Total cost of ownership for the DXP is 0.02 CHF lower than for the ICE

The result of the total cost of ownership modeling can be seen in Fig. 11. Although a savings of 0.02 CHF/km does not seem like an impressive margin, our conservative assumption is that a typical small national postal fleet drives 150,000 km/day (roughly what the Swiss Post is travelling), then over the vehicle fleet's seven year lifetime, we estimate savings of 5,561,729 CHF from switching to SEVs. However, the more significant effect results from the increased payload efficiency of SEVs compared to ICE vehicles. We could estimate roughly a 3× higher payload for SEVs, saving 3× the number of replenishment trips and reduced overall number of vehicles, which leads to substantial cost reductions.

6 Circular Economies

In this final section, we describe our answer to the question: *What can be done at the end of life for SEVs which further increase their economic value and reduce the demand for critical raw materials?*

In 2019, we received 1,153 vehicles back from the Swiss Post as part of a buy-back agreement. To improve the ecological benefit of our vehicles, we created a refurbishment production line that we call KYBURZ SecondLife. To date, we sold 198 refurbished vehicles, and 63 in as-is condition, representing an overall net economic loss, but at substantial ecological gain. The economic benefit of battery recycling is not presently known, but as part of this project KYBURZ has also commissioned a battery recycling line to handle the recycling in-house (Fig. 12).

Details of the scheme and the material balance and flows are shown in Fig. 13. Although the overall economic benefit of the SecondLife project is presently not positive, and the outlook is uncertain, KYBURZ will continue the project since the ecological benefits are clear, such as the advantage of being a family-owned and operated business.

Fig. 12 KYBURZ SecondLife project refurbishment hall and production line

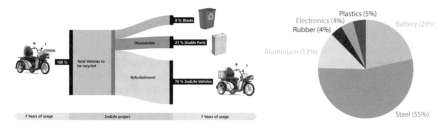

Fig. 13 Majority of the KYBURZ DXPs are refurbished, and those which are recycled result in good yields

7 Summary and Conclusions

To conclude, in this short chapter, we describe our successful deployment of SEVs in the form of the KYBURZ DXP, and draw the following conclusions:

1. Telematics data is extremely valuable for companies in the SEV market wishing to better understand their products,
2. SEVs are at least one third, and perhaps as much as 50% less CO_2 emitting than their ICE-based competitors,
3. The use of SEVs in package delivery operations in urban environments could easily save over 20 km^2 per operation day, reducing congestion and point-source emissions,
4. The total cost of SEV ownership is lower than ICE-based competitors due to reduced replacement rates, greater carrying capacity, and increased operational efficiency,
5. Though it is presently economically irrational, the ecological advantages from refurbishing SEVs for second life deployments means KYBURZ will pursue this approach.

References

1. Allen, J.P.: Understanding the impact of e-commerce on last-mile light goods vehicle activity in urban areas: the case of London. Desirable Trans. Futures (Special Edition), 6–26 (2017)
2. Baeriswyl, M.: Personalzeitung Swiss Post. Retrieved from Die Post (2017). www.post.ch/online-zeitung
3. Cox, B.B.: The environmental burdens of passenger cars: today and tomorrow (2018). Retrieved from PSI Reports: https://www.psi.ch/sites/default/files/import/ta/PublicationTab/Cox_2018.pdf
4. Edwards, J.: Comparative analysis of the carbon footprints of conventional and online retailing: a "last mile" perspective. Int. J. Phys. Distrib. Logist Manage 40 (2010). https://doi.org/10.1108/09600031011018055
5. Kyburz Switzerland AG: DXP 5.0 Brochure (2020). Retrieved from Kyburz Switzerland https://kyburz-switzerland.ch/library/Reduced_Size_Broschuere_Kyburz_DXP_p1-12_7a18.pdf
6. Sacchi, R.B.: Carculator: an open-source tool for prospective environmental and economic life cycle assessment of vehicles. When, Where and How can battery-electric vehicles help reduce greenhouse gas emissions? Renew Sustain Energy Rev, In Review (2019)
7. Williams, G.: Ford Transit tops European and global sales league for cargo vans (2016). Retrieved from Business Vans: https://www.businessvans.co.uk/van-news/ford-transit-tops-european-and-global-sales-league-for-cargo-vans/

BICAR—Urban Light Electric Vehicle

Hans-Jörg Dennig⬛, Adrian Burri, and Philipp Ganz

Abstract This paper describes the technical features of the light electric vehicle (L2e-category) named BICAR. This specially designed vehicle is an all-in-one emissions-free micro-mobility solution providing a cost-effective and sustainable mobility system while supporting the transition towards a low carbon society (smart and sustainable city concept). The BICAR represents part of a multimodal system, complementing public transport with comfort and safety, relieving inner-city congestion and solving the "first and last mile" issue. The BICAR is the lightest and smallest three-wheel vehicle with weather protection. Due to the space-saving design, six to nine BICARS will fit into a single standard parking space. Safety is increased by an elevated driving position and a tilting mechanism when cornering. The BICAR achieves a range of 40–60 km depending on the battery package configuration in urban transport at a speed of 45 km/h. It features a luggage storage place and exchangeable, rechargeable batteries. The BICAR can be driven without a helmet thanks to the safety belt system, which is engineered for street approved tests. The BICAR has an integrated telematic box connected to the vehicle electronics and communicating with the dedicated mobile application, through which the BICAR can be geo-localised, reserved, locked/unlocked and remotely maintained.

Keywords L2e · Shared mobility · Light electric vehicle (LEV) · Cradle to cradle · Solar cell · Urban transport

H.-J. Dennig (✉) · A. Burri · P. Ganz
ZPP Centre for Product and Process Development, ZHAW Zurich
University of Applied Sciences, Winterthur, Switzerland
e-mail: hans-joerg.dennig@zhaw.ch

© The Author(s) 2021
A. Ewert et al. (eds.), *Small Electric Vehicles*,
https://doi.org/10.1007/978-3-030-65843-4_12

1 Introduction

In 2014, a research team led by Adrian Burri and Hans-Jörg Dennig at the Zurich University of Applied Sciences (ZHAW) initiated work on a new vision for an innovative, individual and eco-friendly mobility solution with maximum safety standards and comfort [1]. Two years of research and development activities led to the creation of a pioneering product for urban mobility, the BICAR. After extensive positive feedback from mobility experts, visitors at fairs and end costumers, the ZHAW spin-off company "Share Your BICAR AG" was founded in 2016 with the goal of finalising and commercialising the BICAR. The BICAR is designed for individual users to cover short and medium distances in urban areas. Target customers are mobility sharing providers, business fleet providers and private individuals.

The latest version (see Fig. 1) was presented at the Geneva International Motor Show 2019, generating immense international interest from city authorities, firms, private customers and the media. This article describes the technical features and their background of the current model.

Fig. 1 BICAR—Main features: 1. weather protection incl. wiper system; 2. solar powered; 3. safety belt system; 4. three-wheel-tilting chassis; 5. needs only 1.2 m^2 parking space; 6. robust tubular frame

2 Main Features

The BICAR is the lightest ($\approx$120 kg) and smallest (1.2 m^2 footprint), electrically powered (in-wheel motor) three-wheel vehicle on the market (see Table 1). The BICAR design has been focussed from the beginning on space-efficiency (ease of parking, agility in urban environments) and lightweight construction (maximum energy efficiency). Six to nine BICARs can be parked in a conventional car parking space, relieving road parking space. Therefore, car sharing providers can offer more sharing vehicles, and customers will appreciate that parking space for the BICAR is more easily available than for conventional vehicles. It is sustainable and emissions-free, Cradle to Cradle® certified and energy autonomous with integrated solar cells and, optionally, a backup battery-swap system. The BICAR is an L2e-category vehicle, which can be driven with an AM (EU) driving licence for two- or three-wheel vehicles with a maximum design speed of not more than 45 km per hour. Hence, no car driving licence is needed.

Available in-wheel motors on the market are either too weak (max. 1 kW) or too powerful conceived for heavier vehicles, which makes them too heavy, expensive and inefficient for a lightweight vehicle like the BICAR. In collaboration with an in-wheel motor supplier, an in-wheel motor for vehicles in the weight category <200 kg was developed.

The BICAR is designed for urban use and therefore has a maximum speed of 45 km per hour, which enables construction to be much lighter and thus energy-saving than for micro-cars. The BICAR's energy consumption is even lower than that of a standard motorcycle[1] and up to 95% less than that of a conventional

Table 1 Overview of selected light electric vehicles (LEVs) with weather protection

Model	BICAR BICAR	Renault Twizy	Toyota iRoad	Toyota COMS	Tremola AD Tremola	Torro Velocipedo	Kyburz PLUS
Category	L2e	L7e	L2e	L7e	L6e	ND	L6e
Wheels	3	4	3	4	3	3	4
Speed (km/h)	45	80	45	60	80	80	30
Weight (kg)	120	550	300	408	270	245	250
Footprint (m^2)	1.2	2.87	1.93	2.4	1.6	1.7	ND[a]
Passengers	1	2	2	1	2	2	1

[a]ND = Not defined

---

[1]The energy consumption of light electric vehicles in urban use depends on their weight. The rolling resistances are similar for all vehicles. The aerodynamic resistance plays a minor role in urban traffic.

car [1]. According to the New European Driving Cycle, the BICAR uses 48 Wh/km. The BICAR achieves a range of 40–60 km depending on the battery configuration in urban transport.

A semi-closed cabin construction was designed to protect the driver from wind, rain or snow, while providing enough air circulation to avoid the need for air-conditioning and featuring medium-sized luggage capacity. A three-point safety belt/seat construction connected to the vehicle frame ensures safety for the driver without the need for a helmet.

An integrated telematic box allows communication between the user's phone and the on-board electronics, providing information about vehicle status and access to telemetry data collected through the backend software. These are: battery status (charging and discharging cycles incl. recuperation, battery and environmental temperature, running hours), real-time motor-management-data (ampere and voltage of the motor consumption, vibration, acceleration levels) and usage (geo-localization, driving data, user data). The data is remotely accessible, and maintenance service can be planned before any failure/breakdown occurs (predictive maintenance).

Due to the reduction exclusively to the main functions, design elements in the interior and exterior and thus additional components were avoided. If a vehicle for individual traffic is developed exclusively for urban use, the chassis and structure can be designed simpler. All this leads to significantly lower production costs than for other small three- or four-wheeled vehicles. Maintenance costs are also kept low due to the robust construction and as the telematic box provides predictive maintenance.

3   Cradle to Cradle Production

For the BICAR, a sustainable material- and production-concept according to the Cradle to Cradle principle was developed. Cradle to Cradle® (C2C) design means that all the resources used for the product are reusable in an infinite recycling process. Further, no environmentally toxic materials, liquids or chemicals are used. With the current state of available certified materials, it is possible to obtain Cradle to Cradle "Bronze" certification. This means that at least 75% of all materials used in the BICAR are recyclable. It is furthermore possible to use materials that have a biological cycle, as well as materials that go through a technical cycle. Solutions are available for the steel frame, skin (e.g., polypropylene deep-drawn parts, biological fibre structure plates, etc.), windscreen, solar cells and components for the wheel suspension (steel, aluminium). To date, there is no market solution for Cradle to Cradle tyres (rubber), electronics and batteries. This represents a large field for future research.

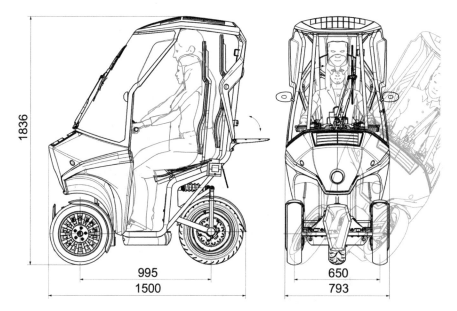

Fig. 2 Layout (numbers in mm) including tilting view and different seat positions (blue: 5% F, green: 95% M)

4 User-Centred Layout Conception

The whole development of the layout is based on a user-centred concept (see Fig. 2) [2]. For this, a tall man (95%M)[2] and a short woman (5%F)[3] were used to modify the ergonomics for the driving position, the location of the steering gear, battery position, interior and exterior design, etc. In order to achieve this, an adjustable seat model was developed that provides an oblique movement in the direction of travel. This allows drivers of different heights to have easy access to the vehicle, to control the vehicle very well when stationary and provides a very good overview while driving. Additionally, the handlebars provide the driver with a firm grip.

---

[2]Tall man: 95th percentile man, man of the body height group 95th percentile man.

[3]Short woman: 5th percentile woman, woman of the body height group 5th percentile woman.

5 Frame and Structure

In order to meet the legal requirements for road approval, the belt anchor points, including the frame, must be subjected to a static load test. The requirements for the test are specified in the 3/2014-EU-Regulations [3].

A force of 6750 ± 200 N acts on the upper belt anchor points and a total force of 6750 ± 200 N also acts on the two lower ones (see Fig. 3). The direction of the acting forces is $\alpha = 10 \pm 5°$ from the horizontal. The frame is firmly clamped at the positions where the front and rear axle chassis is mounted. Here, an additional longitudinal force must be exerted which is ten times the seat weight. The frame is allowed to deform plastically during the test and connections may partially break or crack.

The finite element method (FEM) [4] analysis shows the main load points 1 and 2 (see Fig. 3). S355 steel is used for the main load-bearing tubing and plates. S235 is used for the tubing which is less load bearing; this mainly serves as a supporting structure. At no point is the yield strength or the permissible deformation reached. To verify these calculations, a non-destructive test was carried out with a frame and a clamping system as described in the FEM.

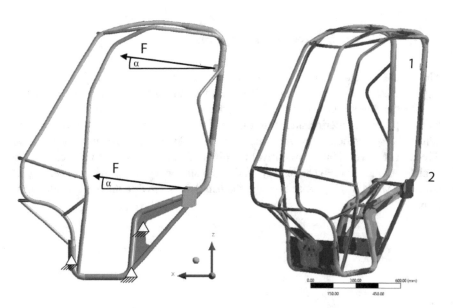

Fig. 3 Structure and stressed areas (*right*)

6 Chassis

During the development of the chassis, special emphasis was placed on a bicycle-like ride behaviour. So, the BICAR tilts when cornering, analogous to the cornering action of a bicycle. This enables fast habituation and ensures safety. The three-wheel construction and the tilting mechanism ensure high manoeuvrability and stability even on slippery road surfaces. For optimum handling, the BICAR has been equipped with a mechanism that allows the vehicle to tilt up to 35° when cornering.

The tilting mechanism has been developed specifically for the BICAR, since existing tilting mechanisms on the market are either unstable and produce unwanted steering forces, making the vehicle difficult to handle, or are far too heavy due to their design for motorcycles with a maximum speed of >120 km/h. The BICAR tilting mechanism is based on a double wishbone suspension with isosceles wishbones and has a very light and cost-efficient construction (see Fig. 4). By connecting the two wheels through a parallelogram formed by the upper and lower wishbones, the wheels always tilt equally. The steering is based on the Ackermann principle and is implemented by means of two track rods. The maximum steering angle of the wheels is 30°. The zero track ensures a secure directional stability in combination with the inclination of the whole vehicle. This allows a very agile and yet stable handling. The horizontal spring-damper unit acting on both sides absorbs axial shocks caused by uneven road surfaces.

The front suspension also includes a tilt brake, which can be activated either automatically or as required when the vehicle is stationary or moving slowly. It is based on a friction-locked principle.

Figure 5 shows the rear wheel suspension in detail. This is a one-sided swing system that enables the motor to be easily replaced in case of damage or for

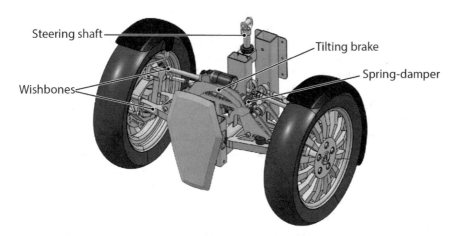

Fig. 4 Isometric views of the front chassis

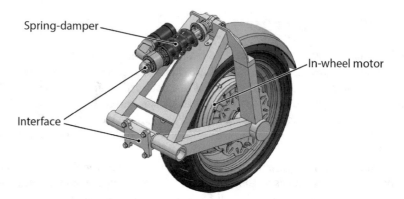

Spring-damper

In-wheel motor

Interface

Fig. 5 Isometric view of the rear suspension with strut (brakes not shown)

maintenance purposes. The rear suspension is attached to the chassis via a lower joint and the spring-damper component. The rear wheel also includes a driving brake and a mandatory parking brake. As with the frame, the chassis structure is a welded construction made of different steels.

7 Energy Management

When considering vehicle sharing solutions, an important criterion for the profitability of the business model is the ubiquitous availability of a vehicle. For electrically powered vehicles, an innovative solution is therefore necessary so that the user can borrow a vehicle with enough range for their desired journey at any time. With the BICAR, several solutions were integrated which have advantages depending on the operating model and location of use.

The battery in the BICAR is exchangeable and installed in the foot compartment. The battery, which weighs less than 8 kg, can be removed by any user, if necessary, and replaced with a charged battery at a battery changing station. Battery changing and charging stations could be set up at defined locations within a city network. The battery could also be replaced by a service team circulating from BICAR to BICAR with charged batteries and replacing the low batteries promptly in time. This service model would be feasible with BICAR used in a free-floating system. This would entail the user being allowed to park the vehicles anywhere within a defined geo-fence (e.g., urban areas or defined parking zones). At these locations, however, charging stations would not be available. In a fixed station-based operating concept, charging points could be installed, so that the vehicle could also be charged via a plug-in power cable.

To this date, the BICAR is the only L2e sharing vehicle on the market with solar cells integrated in the weather protection hood. This surface is large enough to recharge up to a third of the battery capacity. This additional charging facility directly reduces service operations and frequency of battery changes, thus reducing operating costs. The power absorbed by the solar panels depends on the geographical location of the site, season, local weather and the local building architecture (shading of streets). In favourable locations (Southern Europe, South Asia, Central America), the BICAR can be operated all year round as a 100% energy self-sufficient solution, using only solar power.

8 Conclusion

Share your BICAR AG develops and sells new electric mobility solutions optimised for short and medium distances in urban areas. The technology of the BICAR described here is the result of several years of work and technical adjustment after numerous tests with users, customers and experts. Share your BICAR does not act as a sharing provider but as a vehicle manufacturer. For more information: see www.bicar.ch.

At the beginning of the developments, it was not clear whether a vehicle like BICAR with its main functions (see Fig. 1; Chapter "Small Electric Vehicles— Benefits and Drawbacks for Sustainable Urban Development") is satisfactory and drivable from the users' point of view. The construction of several prototypes, the presentation at the Geneva Motor Show 2019 and the subsequent test drives with numerous people have proven this.

The weather protection, the agile and safe driving (belt system, 3-wheel chassis) could be tested in numerous rides. In this context, it was confirmed that especially in urban areas, the use of the BICAR can offer an added value. This article describes the use of the BICAR for sharing providers. Of course, the BICAR can also be used for business fleets or private use.

References

1. Dennig DH-J, Sauter-Servaes DT, Hoppe DM, Burri A (2016) World light electric vehicle summit Barcelona, 20 and 21 September, p 8
2. Mueller A, Maier T, Dennig H-J (2011) Holistic and user-centered layout conception of an electric vehicle (E-Car). Stuttgart, pp 93–106
3. Regulation (EU) No 3/2014. Vehicle functional safety requirements for the approval of two- or three-wheel vehicles and quadricycles, October 24, 2013
4. Steinke, P.: Finite-elemente-methode, 2. Springer, Berlin, Heidelberg, New York (2007)

Conception and Development of a Last Mile Vehicle for Urban Areas

Andreas Höfer, Erhard Esl, Daniel Türk, and Veronika Hüttinger

Abstract In megacities, increasing globalization effects are leading to rapidly increasing prosperity and augmented purchasing power, and thus to a growing need for punctual, cost-effective, and environmentally friendly delivery of goods. A smart, small electric vehicle concept is presented that targets on meeting the requirements for the delivery of goods in urban areas and that is designed especially for the delivery on the last mile. This last mile vehicle (LMV) for cargo transportation is attached to a truck. Whenever it is needed, for example to deliver goods into narrow streets, in pedestrian areas or in case of traffic jams, it can be unfolded and unloaded from the truck and hereby guarantees a flexible and punctual delivery of goods. This flexible on-time delivery is possible because the last mile vehicle is designed, so that the legal regulations of the non-motorized vehicle lane, that is everywhere to be found in Asia, are met. The vehicle is designed with three wheels, a range of 40-60 km and an electric drive train with a continuous power of 2×250 W that enables a maximum speed up to 40 km/h of the vehicle. The drive train consists of a battery pack that can be charged electrically from the truck, two inverters, and two electric wheel hub motors. The LMV has been designed and constructed as a prototype and has been tested on non-public roads to prove the vehicle concept. For Europe, it can be classified as an L2e vehicle and with slight modifications; it can be applied on European roads as well.

Keywords Last mile vehicle · Small electric vehicle · Smart mobility · Connected car

A. Höfer (✉) · E. Esl · D. Türk · V. Hüttinger
Technical University of Munich (TUM), Boltzmannstr. 15,
85748 Garching bei München, Germany
e-mail: hoefer.andy@gmail.com

A. Ewert et al. (eds.), *Small Electric Vehicles*,
https://doi.org/10.1007/978-3-030-65843-4_13

1   Introduction

In China, there is an old saying: "要致富先修路" (loosely translated: If you want to be rich, you must first build roads) [1]. Especially in Chinese megacities, this wisdom has been put into effect for quite some time, leading to prosperity in the ever-increasing middle class and to urban infrastructures that have reached their maximum of capacity in many places. Nevertheless, the urbanization continues (e.g. Shanghai had a population growth of 38% in the years from 2000 to 2011 [2]) and studies predict that by 2030, one billion people will be living in urban areas in China. This will produce a tenfold increase of the traffic volume [3]. The high population density, as well as the population's desire for motorized transport, is leading to the well-known megacity problems such as harmful emissions, smog, and sustained traffic congestion. Electrification, car-sharing, and an expansion of public transportation provide the first solutions for passenger transportation. However, few alternative solutions have been developed so far for the transportation of cargo, although the demand for an on-time delivery of the goods to the final customers continues to grow in importance [4]. Public access to the Internet and the associated introduction of e-commerce accelerates the development of rising freight flows. This brings great challenges to logistics companies especially for the last mile delivery stage [4]. The last mile is the final delivery stage in the logistics chain in which the goods reach (final) customers. Here, the cargo has to be delivered on time in the densest neighborhoods, via narrow one-way streets, into the historic city centers or into pedestrian areas to different customers. Thus, the last mile is not only the bottleneck of the delivery chain, but also provides the largest potential for innovative vehicle concepts and mobility solutions.

2   Methodical Approach

When developing vehicles for cargo transportation, one has to start with the transport task including its customer and market-specific requirements [5]. For this purpose, a market analysis was carried out for the example of Shanghai to understand the logistics chain in Asian megacities (Fig. 1, step 1). The analysis included, among other aspects, expert interviews, traffic flow observations or visits to various logistics centers (details of the market analysis compare [6, 7]). From this, explicit product requirements could be derived such as the length of the last mile or transport volumes and masses (Fig. 1, step 2). Along with the legal regulations as well as the customer's needs, as determined by surveys among truck drivers, a list of requirements for specific vehicle concepts for the last mile is derived. This list of requirements was the basis for subsequent methodical concept development and concept evaluation using product development methods from [6] such as the morphological method (Fig. 1, step 3). The most suitable concept was then further detailed using CAD methods for design and simulation (Fig. 1, step 4). Finally, a

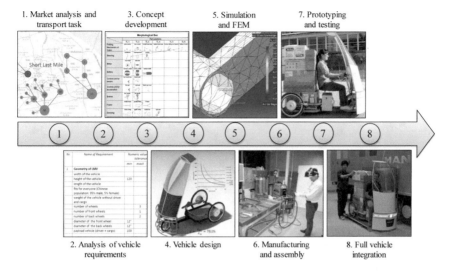

Fig. 1 Product development process for the development of a cargo vehicle for urban areas starting from the definition of the transportation task (step 1) leading to the prototyping and testing (step 7) as well as to the overall vehicle integration (step 8)

prototype of the vehicle concept was manufactured, assembled, and tested (Fig. 1, steps 5–8). A special characteristic of the developed vehicle is its integration into the truck, which greatly affects the system boundary conditions for the design and construction. For example, the ground clearance of the truck has to be kept, which may restrict the available space for stowage of the last mile vehicle (LMV). On the other hand, the integration opened new degrees of freedom such as the connection of the LMV to the 24 V electrical system of the truck.

The applied product development methodology for last mile delivery vehicles is shown in Fig. 1.

3 Field Study, Use Cases, and Vehicle Requirements

In Shanghai, a substantial distinction can be made between four different logistic situations [6]. These can be seen in Fig. 2. Logistic situation 1 (Fig. 2, top left) is the original form of goods transportation in China. Truck drivers, working on a freelance basis with small delivery trucks, wait for orders at collection points near large urban thoroughfares. One delivery journey is to pick up the goods at a place X and deliver them to the final customer C. Afterward, the driver returns to the starting point. This entails low capacity utilization and many empty journeys. The last mile is the total distance traveled by the delivery truck; in this case, it is up to 40 km. The transport of foods, described by logistics situation 2 (Fig. 2, bottom left), is based on a multi-level market system via which the goods are moved to the

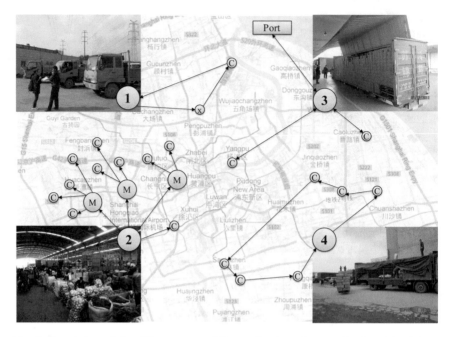

Fig. 2 Description of the different cargo delivery situations that can be found in Chinese megacities today

end customer C. Here, the goods are transported over a last mile that is only up to 2 km. This last mile often includes transportation in pedestrian areas or in the tight confines of the historical city center. Modern transport logistics structures can also be found in Shanghai (Fig. 2, top right and bottom right) and follow the principle *mega hubs in megacities*, which is further described in [4].

In this case, the delivery takes place on optimized routes that are driven along either on a direct route or a star-shaped route, depending on the number of identical goods, volume, and location of the end user. Thus, the distance of the last mile varies between 20 and 80 km depending on type of route. The market analysis yields a required minimum range for the LMV of $R_{min} = 40$ km [6]. This way three of the four logistics situations can be served.

Other important vehicle concept requirements are presented qualitatively in Fig. 3 and are quantified in [7]. The requirements are brought together from six different fields of research. Those include the legal requirements, the requirements to the transportation task (compare Fig. 2) especially on the last mile, urban structures of megacities, the interfaces to the delivery truck, and finally the ergonomic aspects and desires of truck drivers (compare [7]).

Exemplarily two requirements (compare [7]) are further quantified: The need for high vehicle agility results in the requirement of the small turning radius of $r = 1.25$ m. The minimum load capacity in volume should be $V_{load} = 1$ m^3 and $m_{load} = 80$ kg in mass [7].

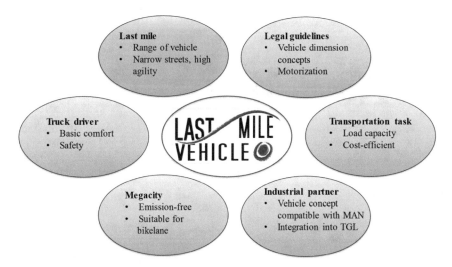

Fig. 3 Key vehicle requirements derived from six fields of research

In addition, the survey carried out with Chinese truck drivers resulted in the need for the provision of basic comforts such as protection against bad weather and the desire for greater active and passive safety measures. A special feature of the LMV is the conceptual design and suitability for bike lanes, which can be found all over metropolitan Shanghai. This hybrid solution increases flexibility for a timelier delivery of goods. The bike lanes are less often affected by traffic jams and traffic obstructions than the main roads. Thus, a flexible on-time delivery of goods is mostly possible. The hybrid design of the LMV for both main roads as well as bike lanes leads to various specific requirements. For example, the requirements for the bike lane bring legal restrictions to the vehicle design such as the number of wheels or the drive concept. Regarding the drive, combustion-powered vehicles are not allowed, whereas electrified vehicles with limited maximum power are permissible.

4   Draft Designs, Evaluation, and Conceptual Component Design

Using diverse creativity techniques from [8] such as brainstorming or the "6-3-5" method, innovative concepts could be developed and systematically structured with morphological boxes. Finally, the concepts were evaluated with a benefit analysis. In Fig. 4, six selected design sketches of vehicle concepts for the last mile that were developed are shown.

Consequently, the six design ideas and their advantages are shortly described. The "Roadrunner" concept is a further development of a Segway™ designed for the

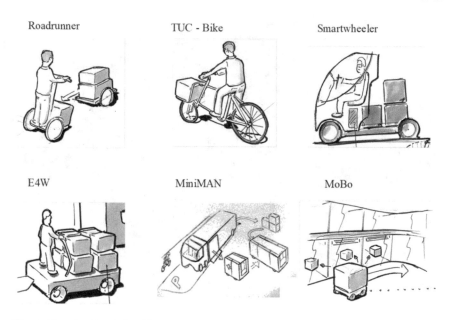

Fig. 4 Visualization of six different small electric vehicle concept for the last mile delivery

transportation of cargo. The vehicle concept shown here is modular and expandable at the same time, allowing multiple units of goods to be transported. The "TUC-Bike" concept is a cargo bike based on a pedelec. In contrast to many state-of-the-art pedelecs (see [8]), a folding mechanism was planned. As a consequence, the bike can be carried on a truck in a space-saving way. The cargo area is extendable and adaptable to the packet size. The "Smart Wheeler" is a four-wheeled electric vehicle that has an enclosed driver cabin. This cabin protects the driver from environmental influences. The cargo is placed behind the driver, and the electric energy storage devices are placed beneath the driver's seat. Thus, a low center of gravity for the vehicle is realized. The "E4W" is considered to be a cuboid vehicle concept that focuses on maximizing payload. The cuboid includes all drive components and the energy storage devices. As a consequence of the compact design, space-saving stowage on the ladder frame beneath the truck is possible. This enables a quick setup of the vehicle and requires no load volume due to the position beneath the superstructure of the truck. The conceptual idea of "MiniMAN" is to divide the loading space of a 7.5t truck into several modular units. On these modular units, the cargo is already pre-consolidated. Furthermore, these modular units contain their own drivetrain and storage components and thus can act as an independent, small electrical supply truck on the last mile into downtown, while the 7.5t truck waits at the edge of town. The "Mobo" concept is related to the "MiniMAN" concept. In contrast, the modular units will drive autonomously. Furthermore, the vehicle units have standardized sizes, such as the size of one euro-pallet. These pallet-sized vehicle concepts deliver the goods to the end customers without a driver and thereby shuttle multiple times between customers and truck.

5 Prototyping

5.1 Vehicle Design and Characteristics

Based on a Chinese tricycle, the MANgo concept was implemented in the form of a functional prototype (Fig. 5a) as it was evaluated the highest in the concept evaluation phase (further details to the value benefit analysis can be found in [7]). Being a purely electrically driven three-wheeled vehicle, it is designed, so that it is suitable for bike lanes in Asia. To be able to drive within urban traffic flow on major roads, the vehicle is equipped with two wheel hub motors with a combined electric peak power of $P_{max} = 2 \times 250$ W and a maximum torque of $M_{max} = 2 \times 40$ Nm. This enables a maximum velocity up to $v_{max} = 40$ km/h. For the wheel hub motors, brushless direct current (BLDC) motors with external rotors are used. The motors are held with spokes directly inside the middle of the wheel (Fig. 5b). The spokes and tires thus have a resilient effect and thereby protect the drive against impacts, which has a positive effect on the lifetime of the electrical machine. The two energy storage units are placed under the driver's seat (Fig. 5c). Their capacity of 0.5 kWh results in a range of 40–60 km for the vehicle. The powertrain has an intentionally redundant structure in order to compensate failures (Fig. 5d). The suspension is designed as a semi-rigid axle (Fig. 5e) and can even ride on curbs. One special technical feature is the attachment of the pushrod actuated air springing/damper units in a space-saving horizontal orientation. The torsional rigidity of the rear axle is designed, so that the wheels can deflect independently from each other. The stator of the wheel hub motor functions as an integrated wheel carrier. Furthermore, Fig. 5c shows the front wheel assembly including a disk break. Figure 5d shows the folding mechanism to convert the vehicle to the folded state [7].

The main technical data of the vehicle is summarized in Table 1 below.

5.2 Key Characteristic—Vehicle Integration

A special feature of the MANgo is the driver cabin that is foldable and protects the driver from environmental influences. Thus, compared to conventional tricycles, the cabin boasts greatly improved comfort. Due to the folding mechanism of the cabin, the vehicle can be stowed in a box carried underneath the body of the truck (Fig. 6; steps 1 and 2).

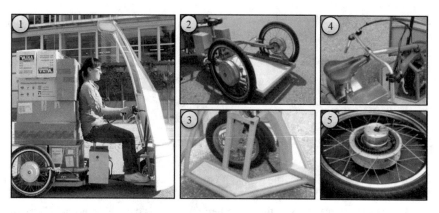

Fig. 5 **1.** Overall vehicle view; **2.** Rear part and cargo area; **3.** Front part vehicle assembly; **4.** Middle vehicle part incl. battery storage and handle bars (folded position); **5.** BLDC wheel hub motor

Table 1 Technical key features of the vehicle concept

Technical key features of the vehicle concept	
Diameter of the wheels d_{front}/d_{back}	13"/19"
Maximum velocity v_{max}	40 km/h
Maximum torque M_{max}	80 Nm
Maximum power P_{max}	500 W
Dimensions in operational state (length × width × height)	1.70 m × 0.76 m × 1.70 m
Dimensions in transportation state (length × width × height)	1.70 m × 0.76 m × 0.55 m
Load capacity V_{load}	1 m³
Maximum vehicle load capacity including driver m_{max}	200 kg
Empty mass m_0	38 kg
Vehicle range R	40–60 km
Ergonomics	5% female, 95% male
Wheel base l	1.22 m
Maximum steering angle δ	35°

This way the MANgo does not subtract any volume from the truck's loading space. The folding mechanism is equipped with two gas springs designed to assist the folding kinematics of the cabin [7].

The setup and the loading of the vehicle for the last mile can be done within 60 s (Fig. 6, step 3). To accomplish this, the MANgo is placed on a platform which also functions as the bottom plate of the storage box. The vehicle can be electrically lowered and raised with the help a spindle drive and two vertically laced linear guides mounted on the ladder frame. Using two horizontal linear guides, the vehicle can be pulled out from beneath the truck and then loaded with goods.

Fig. 6 Truck integration and four step setup process

6 Summary and Outlook

New vehicle and mobility concepts are necessary to meet the increasing challenges of progressive urbanization. At the Technical University of Munich, these challenges were analyzed for cargo transportation and a design methodology for last mile vehicles was developed. For the example of Shanghai, the transportation task for goods delivery in megacities was analyzed, and innovative vehicle concepts were developed and evaluated methodically. The MANgo vehicle concept has been prototypically built and was able to meet all the functional requirements during the testing.

The MANgo that was awarded with a national scientific research award [9] is an electrified tricycle and thus is based on the tradition of Chinese three-wheeled cargo concepts. The foldable driver cabin and the vehicle's compact design allow it to be carried onboard commercial vehicles such as the MAN-TGL. The vehicle also meets the requirements for the use of bike lanes. The implemented technical innovations create a competitive advantage for the vehicle in a variety of different delivery scenarios. For example, in the case of a traffic jam, the MANgo can be unloaded and be set up within a minute. Prioritized goods are transshipped, and on-time delivery is assured. Bike lanes can be used, increasing flexibility and smart mobility. With its electric powertrain, the vehicle is also able to navigate inner cities and deliver into pedestrian areas and is suitable for the narrow streets in old city centers. A more futuristic delivery scenario involves several agile LMVs that are connected with each other and networked with a larger logistics vehicle traveling on outer ring roads. After delivery, the LMVs shuttle back multiple times to the heavy duty truck in order to pick up new loads.

Even though the MANgo has not passed from prototype stage to a serial release yet (2020) that new, small electric vehicle (SEV) concepts including new design methodologies are necessary not only for passenger transport but also for cargo transportation.

The future will show whether small zero-emission vehicles specially designed for the last mile will prevail in logistics in megacities, or if deliveries will continue with conventional large delivery trucks. However, if those conventional delivery trucks are stuck in traffic jams, the vehicle concept presented here for the last mile would still have the advantage of unlimited mobility.

Acknowledgements The scientific content of this paper are resulting from the global Drive Program at Technische Universität München, Lehrstuhl für Fahrzeugtechnik, Fakultät für Maschinenwesen (https://www.mw.tum.de/ftm/lehre/internationale-studentenprojekte/). Some parts were already published in [10].

The results were derived together with Tongji University and MAN Truck & Bus SE. Special thanks to Dr. Britta Michel, Dr.-Ing. Frank Diermeyer, and Alexander Süßmann.

References

1. Biswas, A., Tortajada, C: Bridging the infrastructure deficit. In: The Business Times (2013)
2. Government-China: Infosite Goverment China, 1 July (1999). https://inquiryshanghai.weebly.com/population-characteristics.html, accessed 16 July 2020
3. Textor, M.: (w.Y.) Institut für Pädagogik und Zukunftsforschung (IPZF). http://www.zukunftsentwicklungen.de, accessed 21 January 2015
4. Deutsche Post AG: Delivering Tomorrow—Logistik 2050: Eine Szenariostudie. Deutsche Post AG, Bonn, Deutschland, ISBN 978-3-920269-53-5 (2012)

5. Hoepke, E., Breuer. S.: Nutzfahrzeugtechnik, Köln und Weinheim: Vieweg + Teubner Verlag, Springer Fachmedien Wiesbaden GmbH (2010)
6. Esl, E., Höfer, A., Türk, D., Hüttinger, V.: Abschlusspräsentation: Last Mile Vehicle for Chinese Megacities, München: Technische Universität München, Designskizzen: Bram van Krieken, 24 May (2012)
7. Höfer, A.: Diplomarbeit: Conception and Development of a Last Mile Vehicle: From the Concept to the Prototype. Technische Universität München; Lehrstuhl für Fahrzeugtechnik, München (2012)
8. Lindemann, U.: Methodische Entwicklung technischer Produkte: Methoden flexibel und situationsgerecht anwenden, Garching: Springer. Heidelberg, London, New York, Dordrecht (2009)
9. IAV: Springer Professional: Hermann-Appel Preis für herausragende wissenschaftliche Leistungen in der Fahrzeugtechnik (2014). https://www.springerprofessional.de/automobil—motoren/iav-hermann-appel-preis-2014-verliehen/6585218. Accessed 28 April 2020
10. Höfer, A., Esl, E., Türk, D., et al.: Innovative Fahrzeugkonzepte für Shanghais letzte Meile. ATZ Automobiltech Z **117**, 76–81 (2015). https://doi.org/10.1007/s35148-015-0049-y

Development of the Safe Light Regional Vehicle (SLRV): A Lightweight Vehicle Concept with a Fuel Cell Drivetrain

Michael Kriescher, Sebastian Scheibe, and Tilo Maag

Abstract The safe light regional vehicle (SLRV) concept was developed within the DLR project next-generation car (NGC). NGC SLRV addresses the safety concern of typical L7e vehicles. The SLRV is therefore specifically designed to demonstrate significant improvements to the passive safety of small vehicles. Another important goal of the NGC SLRV concept is to offer solutions to some of the main challenges of electric vehicles: to provide an adequate range and at the same time a reasonable price of the vehicle. In order to address these challenges a major goal of the concept is to minimize the driving resistance of the vehicle, by use of lightweight sandwich structures. A fuel cell drivetrain also helps to keep the overall size and weight of the vehicle low, while still providing sufficient range.

Keywords L7e vehicle concept · Safe sandwich vehicle structure · Fuel cell drivetrain

1 Safe Light Regional Vehicle Concept (SLRV)

The safe light regional vehicle concept (SLRV) has been developed as one of three different vehicle concepts, within the DLR's research project next-generation car (NGC). One important goal of the NGC SLRV concept is to offer solutions to some of the main challenges of electric vehicles: to provide an adequate range and at the same time a reasonable price of the vehicle. In order to address these challenges, a major goal of the concept is to minimize the driving resistance of the vehicle, which leads to a reduced demand for power, energy and resources. The SLRV concept has been developed for regional distances that extend beyond the urban range. It is intended for example, to be used by commuters who have regular point-to-point journeys, such as a feeder vehicle to the public transport connections or as a car-sharing vehicle in an out-of-town context. It can therefore supplement public

M. Kriescher (✉) · S. Scheibe · T. Maag
German Aerospace Center (DLR), Pfaffenwaldring 38-40, 70569 Stuttgart, Germany
e-mail: Michael.Kriescher@dlr.de

© The Author(s) 2021
A. Ewert et al. (eds.), *Small Electric Vehicles*,
https://doi.org/10.1007/978-3-030-65843-4_14

transport in a suburban or rural environment, be used as a second car and is well suited for sharing due to its fast H2 refuelling capability.

NGC SLRV also addresses the safety concern of typical L7e vehicles. The SLRV is therefore specifically designed to demonstrate significant improvements to the passive safety of small vehicles.

The SLRV is a two-seater with a low, elongated body, to provide minimal aerodynamic drag. An innovative metal sandwich structure is used for the car body to keep the vehicle weight low. This allows the use of small, and therefore low cost, drivetrain components, due to secondary weight saving effects [1].

A potential advantage of sandwich structures is the fact that large, relatively simple parts can be designed, without the need for additional stiffeners, reducing the overall number of parts of the car body. This reduction in part count and assembly effort is important to make sandwich structures competitive, since a single sandwich component is typically more expensive than parts made from metal sheets.

An example for this is the passenger compartment of the SLRV which consists of a single floor tray, reinforced by a ring structure. This floor tray substitutes several parts of a conventional car body, such as the door sills, the fire wall and the rear wall as well as the floor itself (see Fig. 1).

In order to reduce the part count further, the SLRV was designed with a single canopy instead of conventional doors. It therefore only needs a rollover bar instead of a roof and A- and C-pillars.

The use of a drive-by-wire steering system simplifies the layout of the car body further, since a steering column with the associated attachment structures becomes

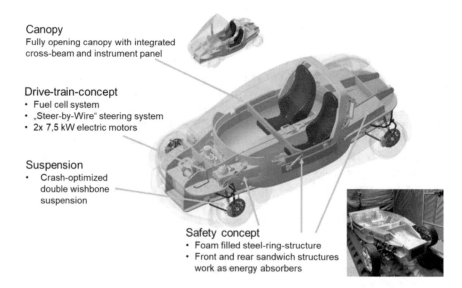

Canopy
Fully opening canopy with integrated
cross-beam and instrument panel

Drive-train-concept
• Fuel cell system
• „Steer-by-Wire" steering system
• 2x 7,5 kW electric motors

Suspension
• Crash-optimized
  double wishbone
  suspension

Safety concept
• Foam filled steel-ring-structure
• Front and rear sandwich structures
  work as energy absorbers

Fig. 1 Overview of the SLRV concept

unnecessary. In the SLRV, the steering wheel is located on a cross-beam which swings open, along with the canopy.

The suspension of the SLRV is optimized with respect to its deformation behaviour during a frontal crash. Several predetermined breaking points are used to avoid an intrusion of the wheel into the cars body (see also chapter Velomobiles and Urban Mobility: Opportunities and Challenges).

2 Concept of the Car Body

An innovative metal sandwich structure is developed to achieve a very low weight for the body in white—only 90 kg—and at the same time optimize the crash behaviour to protect the occupants (see Fig. 2). The use of a metal sandwich structure reduces the number of separate parts necessary for the assembly of the vehicle body. Conventional materials such as aluminium, steel and plastic foam are used to keep material costs low.

Innovative deformation mechanisms are used on several parts of the vehicle body structure, in order to achieve a favourable relationship between crash performance and lightweight design.

The front and rear sections of the vehicle are made from sandwich panels, which are bonded to form the front and rear structures. These structures carry the attachment points for the chassis and most of drivetrain components. They also work as energy absorbers in the case of a frontal or rear impact. The passenger compartment is made of a floor tray with a surrounding ring structure. These components protect the passengers in the case of a side-, or pole impact. They also bear the loads of a front- or rear impact.

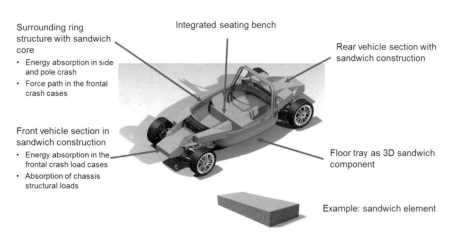

Fig. 2 SLRV vehicle body with metal sandwich construction—weight 90 kg

3   Testing of the Car Body

Two prototypes of the SLRV body in white were built for crash testing. The aim of
these tests was to investigate the deformation behaviour of the entire vehicle body
structure and to verify the anticipated positive attributes of sandwich design, which
had previously already been studied using FE calculations and in the form of
generic components [2].

3.1   Experimental Set-Up and Implementation

Two relevant crash tests were carried out—firstly, a pole crash in line with
EURO-NCAP, and secondly a frontal crash in accordance with US-NCAP.
Investigations into the degree of injury suffered by the occupants could not be
carried out within the context of this project, but the results of the behaviour of the
vehicle structure are an important first result, in order to assess the passive safety of
the vehicle.

3.2   Crash-Test Facility at the DLR Institute of Vehicle
Concepts

The institute of vehicle concepts has a sled system for dynamic tests on larger
components and assemblies. The facility consists of two crash sleds guided by a
system of rails, so that they can only be moved in a longitudinal direction (see
Fig. 3). Sled one, with a total weight of 1300 kg, can be accelerated using a
pneumatic cylinder to a maximum speed of 64 km/h. This allows body assemblies

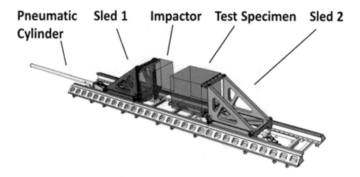

Fig. 3 Arrangement of the crash-test facility at the institute of vehicle concepts [3]

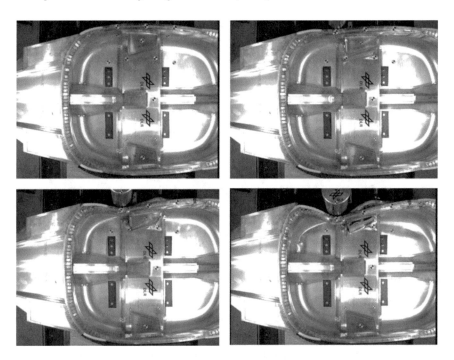

Fig. 4 Deformation behaviour of the body during the pole crash, 0–280 mm intrusion; view from above

for lightweight vehicles to be tested under realistic conditions. The forces during the crash tests (see Figs. 4 and 5) were obtained by measuring the deceleration of the sled.

3.3 Pole Crash Test Implementation

The kinetic energy of the SLRV during the pole crash, at a vehicle mass of 530 kg and an impact velocity of 29 km/h, was 17.2 kJ. Due to the higher weight of the impactor sled compared to the vehicle weight, a slighter lower velocity of 24.4 km/h had to be applied during the crash test in order to achieve the same impact energy. The velocity measured in the test was 24.48 km/h.

During the pole crash, the body exhibited uniform deformation behaviour, without any major reduction in force (see Figs. 4 and 5). At the beginning of the deformation process, there was a good deformation pattern. However, a detachment of the adhesive joints between the floor pan and ring structure, as well as the support for the bench on the ring structure occurred as the deformation continued. This separation of the adhesive joints could not be represented in crash simulations of the pole crash, which leads to future investigation.

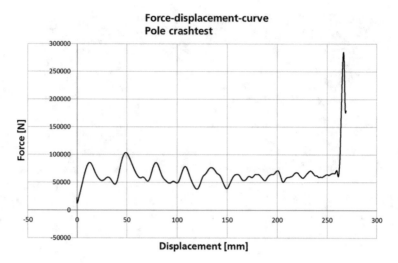

Fig. 5 Force–displacement curve of the pole crash

3.4 Experimental Set-Up for Frontal Crash Test

In a US-NCAP frontal crash test, the vehicle collides with a fixed, non-deformable barrier at 56 km/h [4]. The aim of this test is to investigate the deformation behaviour of the front end in conjunction with the chassis, as well as the structural integrity of the passenger cell.

The vehicle body is firmly connected to sled 2, which remains fixed during the experiment. The barrier, a non-deformable plate, is mounted on sled 1, which is accelerated in the experiment and collides with the fixed body at the set impact velocity (see Fig. 6). The body is connected to fixed sled 2 by a flat support on the rear part, and by plates bolted to the vehicle floor.

The short acceleration distance available for the impactor sled means that high acceleration is necessary to achieve the desired crash energy. In this test, the barrier is accelerated rather than the car to ensure that such acceleration does not lead to a premature deformation of the vehicle body.

As in the case of the pole crash, in this test, the mass of the sled is greater than that of the SLRV, so the impact velocity had to be reduced from 56 km/h to 44.85 km/h in order to achieve the same impact energy.

4 Results of the Front Crash Test

Overall, the SLRV front structure displayed an even, continuous deformation behaviour with sufficient energy absorption (see Fig. 7). The adhesive joints also performed as well as in the simulation. The front suspension worked as planned during the crash and successfully prevented an impact of the wheels on the passenger compartment.

Fig. 6 Left: Experimental set-up for the frontal crash test, with sled 2 fixed; Right: sled 1 in motion

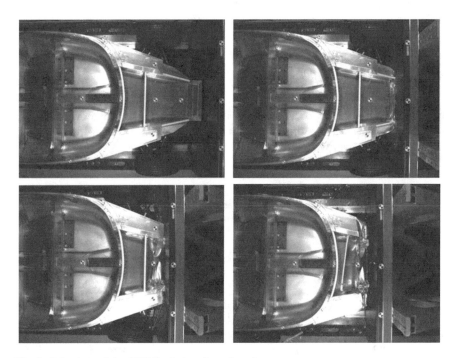

Fig. 7 Behaviour of the *SLRV body in a frontal* crash test

4.1 Behaviour of the Passenger Compartment in the Frontal Crash Test

A slight deformation of the tunnel occurred at the point at which the wheels were impacted, at around 200 mm. Otherwise, no plastic deformation of the passenger compartment occurred during the entire crash test. So the survival space for the passengers remained fully intact (see Fig. 9). Visible elastic deformations occurred in the area of the ring, but they completely disappeared by the end of the test. Disregarding the impact of the wheels, the deformation force of front structure is around 100–120 kN (see Fig. 8). This equals a deceleration of 20–23 g and is therefore below the maximum deceleration of state of the art passenger cars. Therefore, with a working passenger restraint system, a low risk of injury is to be expected.

For the future development of small electric vehicles, an investigation of the occupant's safety, by using crash-test dummies, would be beneficial. This could first be done by simulation and could lead to a more complete test program, including crash-test dummies and the appropriate restraint systems.

4.2 Drivetrain of the SLRV

NGC SLRV is designed for an electric drivetrain, powered by a hydrogen fuel cell system (see Fig. 10). For the targeted range of 400 km, a fuel cell system can achieve a much lower weight than an equivalent battery system [5]. Due to the low driving resistance of the vehicle, the fuel cell system is designed with a low power

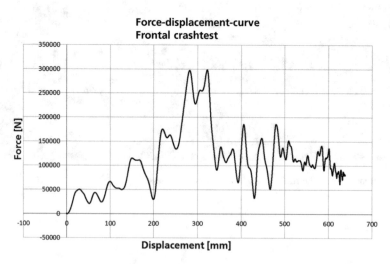

Fig. 8 Force–displacement curve of the frontal crash test with the SLRV body in white

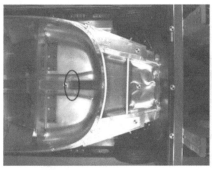

Fig. 9 Appearance of a slight deformation in the tunnel area

output, which limits the cost of the system, as well as the consumption of hydrogen. The challenges are to balance good drivability with the overall weight.

In order to achieve sufficient acceleration, the drivetrain is designed as a hybrid system. It consists of a fuel cell which provides a maximum power of 8 kW and a battery system, which can deliver up to 25 kW, additionally. This limits the cost and weight for the fuel cell and also enables recuperation.

Hydrogen is stored at up to 700 bar, in a single pressure tank, located in the tunnel of the SLRV. The tank has a capacity of 1.6 kg of H_2 at 700 bar. With an estimated fuel consumption of 0.34 kg H_2 for 100 km in the NEFZ cycle, this should theoretically be enough for 470 km range. Since the SLRV is an L-category vehicle, the WLTP C2 cycle would lead to an even lower fuel consumption and an even greater range.

It is understood that small electric vehicles are very cost sensitive (see also chapter 'Small electric vehicles—benefits and drawbacks'). Economic efficiency calculations for hydrogen propulsion have to take the future use of regional vehicles into account. The degree of utilization, the milage per day, as well as hydrogen availability and costs play an important role. This part of the study has not been finished so far and is subject to in future investigations and publications.

4.3 Installation of the Operable Research Vehicle

A research vehicle of the SLRV is being built based on the work described above (Fig. 11). It will be completed and tested until the end of 2020. The goal is to evaluate the concept as well as the performance of all of its systems during test-driving. The tests will evaluate the performance, H2-consumption and diving dynamics of the vehicle. Also, the mechanical loads and the strain on structural components will be measured, in order to evaluate the performance of the sandwich structure under driving conditions.

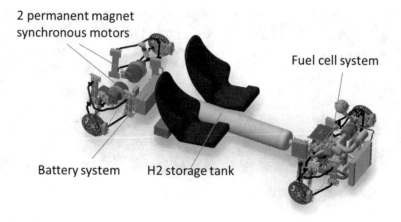

Fig. 10 Drivetrain concept of the SLRV

Fig. 11 SLRV research vehicle, November 2019, the drivetrain is still being developed

References

1. Eckstein, L., Göbbels, R., Goede, M, et al.: Analyse sekundärer Gewichtseinsparpotenziale in Kraftfahrzeugen, ATZ 01/2011 (2011)
2. Kriescher, M., Hampel, M., Grünheid, T., Brückmann, S.: Entwicklung einer leichten, funktionsintegrierten Karosserie in Metall-Sandwich-Bauweise, In: Lightweight Design (2015)

3. Kriescher, M. et al.: Dynamische Komponenten-Prüfanlage, In: handout of the Institute of Vehicle Concepts (2013)
4. Crash Training safety wissen App., 21 March 2018
5. Friedrich, A.: Batterie oder Brennstoffzelle—was bewegt uns in Zukunft? Talk, DLR Institute of Technical Thermodynamics (2014)

Printed in the United States
by Baker & Taylor Publisher Services